Developing Critical Thinking through Science

Book One

June Main

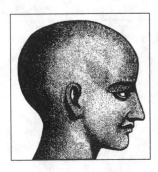

© 1991
CRITICAL THINKING BOOKS & SOFTWARE
www.criticalthinking.com
P.O. Box 448 • Pacific Grove • CA 93950-0448
Phone 800-458-4849 • FAX 831-393-3277
ISBN 0-89455-424-7
Printed in the United States of America

ABOUT THE AUTHOR

DR. JUNE MAIN, professor of education in the School of Education at Jacksonville University, has taught science for elementary teachers at the university level for many years. She previously taught science at the University of Florida and the University of North Florida. Prior to this, she taught in public school for thirteen years at the elementary level.

She has a B.A. and an M.Ed. in Elementary Education and an M.Ed. in Administration and Supervision from the University of North Florida. She holds a Ph.D. in Curriculum and Instruction with emphasis in Science Education from the University of Florida and has worked closely with Dr. Mary Budd Rowe.

Dr. Main is a frequent presenter at regional, national, and international conferences on activity-based, constructivist science teaching and the integration of curriculum with science as a catalyst, always emphasizing strategies to develop children's higher order and critical thinking. She frequently gives regional workshops on these topics. She has been a science consultant and presenter for the Association of International Schools in Africa, recently presenting several workshops at annual teacher conferences in Kenya and Namibia.

Dr. Main is the recipient of numerous awards and honors for excellence in science teaching, including Who's Who in American Education, the National Science Teachers Association's STAR Award, and the international Innovative Excellence in Teaching, Learning and Technology award. She is the author of many books, curriculum materials, and articles in the field of education.

CONTENTS

INTRODUCTION

Man's mind, once stretched by a new idea, never regains its original dimensions.
— Oliver Wendell Holmes

DEVELOPING CRITICAL THINKING THROUGH SCIENCE is based on the premise that students apply thinking skills to learning science concepts and principles by:

Doing through direct, firsthand experiences in an interactive, open atmosphere;

Constructing by building their knowledge through guided inquiry;

Connecting by relating their learning to the world around them.

To help students learn science, four major themes are promoted in this book.

1.

Science can and should motivate students toward learning and toward developing curiosity about the world in which they live.

The contents of this book have been designed to provide teachers and prospective teachers with a variety of science activities that spark this curiosity in students. All of the activities have been tested in classrooms. Each activity includes step-by-step procedures and questioning strategies that help students practice critical thinking while applying it to what they learn in class and to their real world.

These activities can be used successfully with a minimum of science knowledge, preparation time, and science equipment. Most of the materials and equipment are everyday, inexpensive items that can be found around school, home, outside in the environment, or in local stores. Many items can be collected and contributed by the students, thus reinforcing their concept of direct participation in the science activities.

2.

Science is viewed as an active process of developing ideas, or "storybuilding," rather than as static bodies of already-existing knowledge to be passed on to students.

As teachers, we want our students to develop valid ideas about science, and we want the ideas developed in a spirit of inquiry and investigation. This means that instead of merely describing what is taking place, the teacher guides the students through an inquiry process by asking pertinent, open-ended questions and by encouraging investigative process through demonstration, hands-on opportunities, and extension of experiments. Using this process, students can build and comprehend ideas for themselves.

The storybuilding components come, then, from the students' own direct observations, previous knowledge, and personal involvement in the activities. Together with the teacher's guidance, this leads to comprehension of meaningful and valid science concepts and principles.

3.

Students are encouraged to observe and describe their observations accurately and completely.

All student observations and descriptions are considered acceptable no matter how they are stated, for it is from these observations that the information is examined and evaluated for relevancy and organized to build their concept "stories." Students should be encouraged to articulate their observations and ideas using scientific terminology. Scientific terms are defined, demonstrated with concrete examples, then applied and reinforced throughout the activities.

Let's illustrate this process with a model lesson in which students are studying steam condensing on a cool surface. Questions can be added to or deleted from this process as the student's abilities or pre-knowledge dictates.

Teacher *How did we make the steam?*
Students *By heating the water.*
Teacher *When we see steam, we are really looking at water evaporating or escaping into the air.*

Teacher	*What do you think will happen when I hold this mirror over the spout?*
Students	*The mirror will get hot.*
Teacher	*That's true. That's why I'm going to hold it with the oven mitts to protect my hands.*
Teacher	*What do you see on the surface of the mirror?*
Students	*It's clouding up. There are drops of water on the mirror.*
Teacher	*Where did the drops of water come from?*
Students	*The water came from the steam.*
Teacher	*Where does the steam come from?*
Students	*From inside the kettle. From the hot water inside the kettle.*
Teacher	*So what is steam made from?*
Students	*Steam is made from drops of hot water.*
Teacher	*We call the drops of water gathering on the mirror condensation. Let's see how big the drops of water are in steam. I'll dry off the mirror and let it cloud up again. Look at the surface of the mirror. What do you see?*
Students	*Lots of very tiny drops of water.*
Teacher	*Let's collect more condensation. What do you see happening now?*
Students	*The drops are getting bigger. Some drops are running down the mirror and falling off.*
Teacher	*What is happening to make the drops bigger?*
Students	*The little drops are flowing together.*
Teacher	*Why are the drops running down the mirror and falling off?*
Students	*The drops are too heavy. They are falling in a stream.*

Notice the use of terms such as *surface, evaporating, condensing, flowing,* and *falling in a stream.* These are terms students learned as they were "storybuilding" in previous lessons in the water unit. If students have problems recall-ing or supplying these terms you, as their teacher, can model such terms for them in your discussions. You can continue to reinforce, connect, and expand students' use and comprehension of these terms as you demonstrate other related activities or see opportunities while studying other subjects. An example of this would be for students to describe what they see in the illustrations in their reading or social studies books. Guide students to use as many descriptive words as possible.

Ask students to rephrase their observations and descriptions of science experiences so they express themselves more "scientifically." Being able to practice observing and describing in this way not only expands their understanding of science concepts, it enriches their language expression and actively involves them in their own learning.

4.

An open, interactive atmosphere in the classroom is essential.

In an open and interactive atmosphere, the teacher and the class, or small groups of students, actively investigate ideas together (compared to a passive learning situation in which students are merely told the problem, given the answers, and expected to memorize the information). The open-classroom atmosphere removes students' fear of giving a "wrong" answer or explanation and replaces it with a less inhibited ability to express what they think is possible based on their experiences at that time.

The open, interactive atmosphere promotes questioning between teacher/student, student/student, and student/teacher. It encourages prediction, experimentation, and discovery of new dimensions of concepts. It allows examination of "What if?" questions about possible changes in the activity, then allows testing for the results of these changes or *variables.* Activities can be followed as written or expanded. If time allows, the teacher can encourage free expression of students' related curiosities and personal interests and extend an activity by exploring, experimenting, and discussing until all are satisfied. Also, some activities can be repeated for reinforcement using different variables suggested by students.

By becoming involved in these activities through direct observation, hands-on participation, and verbalization of the physical and thought processes, students build a more concrete understanding of the concepts taught in the activities.

They are able to relate and apply these concepts to their environment on a more scientific level. With the teacher's help, students can learn to apply these same analytic and problem-solving skills to their other studies and to any classroom or social problems that might arise.

UNIT 1: OBSERVING

Our contact with the world is through our five senses. We gather basic information in our everyday experiences by seeing, hearing, touching, smelling, and tasting. This process of gathering information by using our senses is called *observing*.

The activities in this unit are designed to guide students to become more aware when using their senses; to make observations using their senses; and to make inferences based on these observations.

ACTIVITY 1: WHAT MAKES A GROUP?

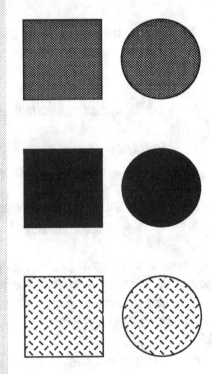

Goal: To understand that objects can be grouped and that there are different ways to group objects

Skills: Observing, comparing, describing, classifying, recording

Materials: Construction paper – blue and red
Sandpaper
Scissors
Sheets of newsprint or white paper
Crayons or markers – blue, red, and brown or tan

Preparation: Cut circles (about 2" in diameter) and squares (with sides about 2" long) so there is a set of shapes:
1 red circle
1 red square
1 blue circle
1 blue square
1 sandpaper circle
1 sandpaper square

Preparation Time: 15–20 minutes

Lesson Time: 30–35 minutes

— Procedure and Questioning Strategy —

This activity works best when done with groups of four or five students. Mix up the circles and squares and put them in a pile on the table.

1. Do these things make a group?
    (Students usually answer yes.)

2. Yes, they are a group. A group is two or more things that are together.

Pick up any shape from the pile and hold it up.

3. Does this one thing make a group?

    No.

4. Why do you think it isn't a group?

    It's only one thing.

5. It does take more than one thing to make a group. So this one thing is not a group.

Spread out all of the colored shapes in the pile.

6. How many things are in this group?

    Six.

7. Look carefully at all of the things in the group. How can we make smaller groups that are alike in some way?

    (Students usually mention color first.)

Have a student put the shapes into the suggested groups.

8. How many smaller groups do we have?

    Three (blue, red, and brown).

9. How are the things in each group the same?

    They're the same color.

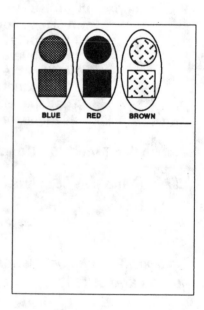

Draw the shapes in each color group on a sheet of newsprint with a crayon or marker of that same color. Leave enough room on the sheet for two other groupings below this one. Draw a line under this first grouping.

10. Is there another way we could group them?

    (Shape is usually mentioned next.)

When this is suggested, ask a student to arrange the shapes into the groups.

11. How many groups do we have now?

    Two.

12. How are the things in each group the same?

    They're the same shape.

13. Is the color the same in each of these two groups?

    No, each piece is a different color.

Record these groups on the newsprint. Draw a line under this second grouping.

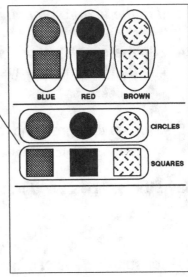

14. How else could the things be grouped?
    (Students usually group them by texture last.)

Have a student group the shapes.

15. Tell us why you grouped them like that.
    (Students usually describe how the things in each group feel, like bumpy and smooth.)

Record these two groups on the newsprint. Hold up the newsprint so the students can see it. Discuss how these groups were made.

16. How many groups do we have now?
    Two.

17. How many items are in each group?
    Two in one group, four in the other.

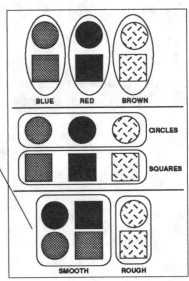

18. How something feels—rough or bumpy like the sandpaper circle or smooth like the red circle—is called its texture.

19. Which group has the most members / the least members?
    The smooth group. / The rough or bumpy group.

Put all of the shapes back into one group.

20. We started with one group—with six things in it.

21. Let's think about the different ways we put them into smaller groups. What was the first way?
    Color.

Point to the color grouping on the newsprint.

22. What was the second way?
    Shape.

Point to the shape grouping on the newsprint.

23. How did we group them the last time?
    Texture.

Point to the texture grouping on the newsprint.

24. So there isn't just one way to group things. There are many different ways we can group them. How many ways did we find to group these six things?

    Three.

25. Does every group have to have the same number of items in it?

    No. Some groups have more than others.

Save the shapes used in this activity for a later lesson.

— Practical Application —

Ask for suggestions of ways that students in the class could be grouped, for instance, by hair color or eye color. Have the students actually form the suggested groups.

* * * * *

Gather a group of small classroom objects with some similar properties. Discuss the similarities and differences of the objects. Ask for ways that these objects could be grouped.

ACTIVITY 2: WHAT IS THAT SOUND?

Goal: To understand that we can identify the source of sound by listening

Skills: Observing by listening, identifying, classifying, describing, explaining

Materials: Audio tape of sounds around the classroom. Suggested sounds:
  closing a door
  walking across the room
  closing a book
  writing on paper
  writing on the chalkboard
  washing hands
  taking a paper towel from the dispenser
  drying hands with the paper towel
  putting a book on the table
  putting a pencil on the desk
  cutting paper with a paper cutter

Preparation: Make the tape of classroom sounds.

Preparation Time: About 15 minutes

Lesson Time: 25–30 minutes plus time over the next few days to play "Knock, Knock."

— Procedure and Questioning Strategy —

1. Let's sit quietly, close our eyes, and listen very carefully.

Give students 2–5 minutes to listen to noises in the classroom and coming from outside. Have them open their eyes and discuss the different sounds they heard and what they think made the sounds. Then ask the following questions.

2. How did you know what was happening when your eyes were closed?
   I heard the sound(s).

3. What part of your body helped you know what was happening?
   My ears.

4. Let's try this! Close your eyes again. I'm going to make a sound with part of my body. Listen and try to guess what I'm doing.

Snap your fingers.

5. What do you think made that noise?
   You snapped your fingers.

Snap your fingers again.

6. How would you describe the noise?

    It's a snapping noise.

7. Is it loud or soft?

    It's loud/soft (depending upon how loudly you snapped).

8. Now keep your eyes closed until I tell you to open them. I am going to make more sounds. I will ask you to tell me what each sound is after I make it.

Let students identify what you are doing each time, then make the same sound themselves— with their eyes still closed. Suggested sounds:

> *tapping foot*
>
> *snapping fingers*
>
> *whistling*
>
> *clapping hands*
>
> *stamping foot*
>
> *rubbing hands together*
>
> *smacking lips*
>
> *humming*
>
> *clicking with tongue*
>
> *repeat some sounds already made in a particular sequence or pattern.*

9. How did you know what I was doing?

    I heard the sound.

10. Let's play another guessing game! I'm going to play a tape of sounds we hear around the classroom. See if you can tell what is making the sounds.

Play the tape, stopping after each sound so students can identify the sound. If students cannot identify a sound, tell them they'll hear that sound on the tape tomorrow. This gives them time to listen to sounds in the classroom and be able to identify the sound on the tape.

11. What part of your body helped you know what was making the sound on the tape?

    My ears.

12. Let's talk about the kinds of sounds you heard.

Encourage students to use words like snapping, clapping, clicking, etc., to describe all the sounds they heard.

— Practical Application —

Over the next couple of days, play "Knock, Knock, Who's There?" One student sits in a chair with his/her back to the rest of the students. Have one student at a time come quietly up to the back of the chair and knock on the back of the chair. When the student in the chair says, "Who's there?" the student knocking responds, "Can you hear who I am?" or something similar.

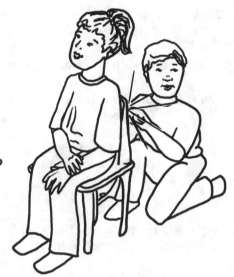

1. How do we know who is talking without seeing them?

   We hear their voices.

2. How can we tell whose voice we hear?

   The voices sound different.

* * * * *

Repeat some sounds in a pattern or sequence for students to identify. Ask students to group sounds they can make such as clicking, tapping, smacking, etc.

* * * * *

Audio tape the voice of each student in the class. You might ask each student to tell a favorite thing to do. Keep a list of the order of the students on the tape. Play the tape over a period of time so students can identify the voices.

* * * * *

Audio tape sounds around the house for students to identify, for example: telephone ringing, washing machine, water going down the drain, water running in the shower/bathtub, car starting, TV playing, clock ticking, radio playing.

ACTIVITY 3: HIGH AND LOW, LOUD AND SOFT SOUNDS

Goal: To differentiate between high and low, loud and soft sounds
To understand that the longer the straw, the lower the pitch

Skills: Observing by listening, comparing, classifying, inferring, explaining

Materials: Drinking straws
Scissors
Cardboard tube from a toilet paper roll
5" square of waxed paper
Rubber band
For each student:
one of each of the items above (Have the students bring in the cardboard tubes.)

Preparation: 1. Gather the materials and place them on a table so they can be easily distributed.
2. Practice making sounds by blowing across a straw, as in the illustration below.

Preparation Time: 10 minutes

Lesson Time: 20–25 minutes

— Procedure and Questioning Strategy —

Show the students how to make a sound with a straw. Hold the straw with your thumb covering the bottom of the straw. Put the straw close to your bottom lip, as illustrated below, and blow across the straw. Blow softly for a soft sound. Blow harder for a louder sound. (You may need to practice before demonstrating for the students.) Now distribute one straw to each student.

1. See if you can make a sound with your straw by blowing across the top of it.

2. Try to make a soft sound as you blow across your straw.

3. How did you do that?
   I didn't blow very hard.

4. Try to make a loud sound.

5. How did you make it louder?
   I blew harder.

Distribute a pair of scissors to each student.

6. Cut your straw into two pieces—a long piece and a short piece.

7. Blow across the top of each piece.

8. How are the sounds different?

    One sound is higher (or lower) than the other.

9. Which straw makes the higher sound?

    The shorter straw.

Cut three lengths of straw—1", 3", and 5"—and hold them up so students can see them. 1"   3"   5"

10. Here are three straws of different lengths. Which one do you think will make the lowest sound?

    The longest one.

11. Let's listen to the sounds.

Have three students, one at a time, blow across each of these lengths of straw.

12. Which one made the lowest sound?

    The longest one.

13. Which one made the highest sound?

    The shortest one.

Hold up the 3" length of straw.

14. What sound did this one make?

    Higher than the longest straw. Lower than the shortest straw.

If the students do not recognize that this straw made a sound in between the other two sounds, have the three students blow over their straws in order, from lowest to highest sounds.

* * * * *

If the students can identify the three different pitches, cut six lengths of straw, from 1" to 6". Select six students to blow across the straws. Give them the straws in random order of size. Have the rest of the students listen to the six sounds and decide how to place them in order from lowest to highest.

— Practical Application —

Demonstrate how to make a kazoo:
- *Hold a square of waxed paper over the open end of a cardboard tube.*
- *Secure the waxed paper with a rubber band.*
- *Poke a hole in the side of the tube, near the end with the waxed paper, with a sharp pencil point.*
- *Hum or toot through the open end of the kazoo, using varying amounts of air force and pitch until you get the right vibration.*

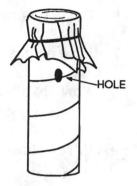

Distribute the kazoo materials so students can make and test their own kazoos. Can they make soft sounds? Loud sounds? Can they make high sounds? Low sounds? (Pitch can be varied by humming higher or lower sounds into the kazoo.) Have the students try humming a tune on their kazoos together.

* * * * *

For further exploration, invite students studying an instrument in school or at home to bring their instrument to class and demonstrate. You might also invite the school orchestra or band—or some of their members—to come visit.

ACTIVITY 4: WHAT DOES IT SMELL LIKE?

Goal: To understand that some objects can be identified by smelling them

Skills: Observing by smelling, comparing, identifying

Materials: Objects from the room that have a particular odor, for example:
soap
crayon
modeling clay
fish food
pencil
chalk
glue
paste
tempera paint
a fresh ditto
Cardboard box with cover (large enough to hold the items above)
Blindfold, one for each student

Preparation: Put the objects in the box. Put the cover on the box.

Preparation Time: 5 minutes

Lesson Time: 10–15 minutes

— Procedure and Questioning Strategy —

This activity is best done with small groups of students. Place the box of objects to be smelled on a table near you.

1. We're going to play a game. Inside this box are things you see in this room every day. Let's see if we can find out what they are just by smelling them. After I blindfold you, I will hold an object near your nose. Don't try to touch it. Try to identify what it is by how it smells.

Let each student, blindfolded, have a chance to smell an object without telling what it is until everyone has smelled it (this way they can compare one smell to another). Remind students that the game is to try to identify the objects without touching them. If a student cannot name an object, give clues about how the object is used in the classroom.

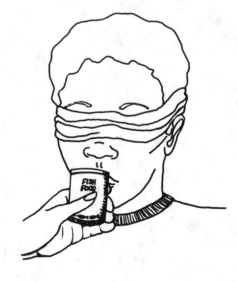

2. What did you do to help you identify something?
    Smelled it.

3. What part of your body did you use to smell?
    My nose.

4. Let's discuss what kinds of smells you smelled. Did anyone smell something sweet or something sour? Something like alcohol or like wood? Were most of the smells pleasant? Were they smells you liked? How many of you smelled something unpleasant, smells you did not like?

— Practical Application —

1. We have found that we can use our noses to tell what something is by smelling it. Now we're going to take a walk in our imaginations. Sit in a comfortable position. Close your eyes. Be very quiet so each of us can discover where we are.

Slowly read the following imagery, pausing at each ellipsis, so students have plenty of time to visualize.

2. Imagine that you are walking along very slowly . . . You stop outside a door . . . You turn the door handle . . . You go in through the door . . . You smell something . . . Take a deep breath . . . What does it smell like? . . . What do you think it is? . . . Can you see it? . . . Where are you? . . . How do you know? . . . Now turn around . . . Go back through the door . . . Close it . . . Walk slowly back to the classroom . . . You are back in the classroom . . . Slowly open your eyes.

Give the students time to think about their experiences. Ask volunteers to describe what they smelled, where they were, why that smell was there, what it smelled like (encourage words other than good or bad).

© 1991 CRITICAL THINKING PRESS & SOFTWARE • P.O. Box 448, Pacific Grove, CA 93950

ACTIVITY 5: TASTING PARTY

Goal: To understand that we recognize foods by tasting them with the tongue, feeling them with parts of the mouth, and listening as we chew them

Skills: Observing by tasting, feeling, and listening; comparing, describing, classifying, inferring, explaining

Materials: Various foods for a tasting party. For example: apples, carrots, banana, grapes, celery, crackers, toast, bread, peanut butter, cheese, miniature marshmallows
Plastic knife
Plastic spoons
Toothpicks
Paper plates or paper towels
Blindfold, one for each student
Paper towel for each student (in case they don't want to swallow some of the foods)

Preparation:
1. Send a note to parents/guardians far enough in advance of tasting activities to find out if any student is allergic to certain foods. Be sure any student participating in this activity has returned a proper response.
2. Cut the foods into small pieces and put them on paper plates or towels.
3. Put the food on a table. Place toothpicks and spoons near the food.

Preparation Time: 5 minutes

Lesson Time: 20–25 minutes

— Procedure and Questioning Strategy —

Since this activity is best done with small groups, divide the class into groups of 6–8 students. Let each group look at all the foods on the table and try to identify them visually. Then, taking one group at a time, blindfold them (or ask students in the group to close their eyes tightly) and ask them to open their mouths. Place a piece of food on their tongues. Tell the students to chew slowly and try to identify their food.

Use a different food for each group. Also use a separate spoon or toothpick for each student. When all groups have tasted and identified their food, help them to discuss what they've done.

1. Let's discuss the ways we've just used to identify the foods we tasted.

2. How were you able to identify the food in your mouth when you could not see it?

By tasting it.

3. What part of your body did you use to taste the food?

My tongue. My mouth.

4. Was there any other way you used to identify the food?

By feeling it in my mouth.

5. What part of your mouth helped you find out what the food felt like?

My tongue. The inside of my whole mouth.

6. So, we've found we can tell what we're eating by tasting it with our tongues and feeling it with parts of our mouths. Let's find out if there is any other way.

Ask for a volunteer to come up to be blindfolded or to close his/her eyes tight. Give this student a piece of celery, carrot, or other crunchy food.

7. What do you think you are eating?

(Students' responses will vary.)

8. Are you using a different way to identify the food this time—a way we haven't discussed before?

(Students usually say they can hear it or feel it crunch.)

9. So sometimes we can use the sound food makes when it is being chewed to help us identify it?

Yes.

Now give each student in the class a paper plate. Advise students how to handle foods, toothpicks, and spoons in a safe, sanitary way. Tell them to touch only the foods they will put on their plates. Tell them to select one food from each of the plates using a toothpick and to spoon out a little of the peanut butter with a plastic spoon, then return to their seats.

10. Put pieces of food in your mouth, one at a time, and chew them slowly and carefully. Think about what each food smells like, tastes like, feels like, and sounds like. Notice how one food is different from another.

Answers will vary for questions 11–16.

11. Which food tasted the sweetest?

12. Which food made the crunchiest noise?

13. Which food felt the smoothest?

14. Which food felt the stickiest?

15. Which food was the juiciest?

16. Which food had the strongest smell?

17. We've talked about three ways we can identify what we're eating? What are these three ways?

    Tasting it with our tongues. Feeling it with our mouths. Listening to the sound as we chew.

— Practical Application —

1. How can we group the foods we've been tasting?

    (Answers will vary including taste, sweetness, color, sound, smell, feeling, texture, etc.)

Have the students group and subgroup any foods that are left on the table using the attributes they generated .

ACTIVITY 6: TASTE THESE!

Goal: To understand that some foods look similar but taste different
To experience that some things cannot be identified by just looking at them or just tasting them

Skills: Observing by seeing, tasting, and smelling; comparing, describing, classifying, inferring, explaining

Materials: The following pairs of foods:
  salt–granulated sugar / flour–powdered sugar
  (small pieces of the foods below)
  apple–cucumber / carrot–sweet potato
  apple–white potato
  apple–white radish or jicama
Toothpicks (heavy or cocktail picks)
Plastic spoons
Small paper plates
Paper towel for each student (in case they don't want to swallow some of the foods)
Peppermint candies, wrapped separately, one per student (in a paper bag so they can't be seen)

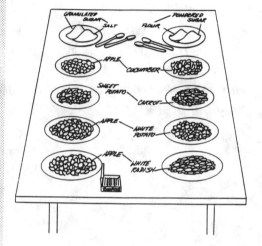

Preparation:
1. Send a note to parents/guardians far enough in advance of tasting activities to find out if any student is allergic to certain foods. Be sure any student participating in this activity has returned a proper response.
2. Pour small mounds of salt and granulated sugar on a paper plate. Pour mounds of flour and powdered sugar on another plate.
3. Peel foods so they are less identifiable (the cucumber should also be cut in half lengthwise and the seeds and soft pulp scraped out).
4. Cut all the foods into pieces, making certain there is one piece of each type of food for each student, plus a few extras.
5. Put each food on its own plate.
6. Put a peppermint candy in your pocket.

Preparation Time: 5 minutes

Lesson Time: 20–25 minutes

— Procedure and Questioning Strategy —

Divide the students into six groups. Give each student a paper plate, a few toothpicks, a plastic spoon, and a paper towel. Now give each group one pair of foods (e.g., salt–granulated sugar, carrot–sweet potato, etc.). Each group will try to identify the foods they have been given.

1. Look carefully at your two foods. What is similar about them?

    They are both the same color. They look just alike. They both look like powder.

2. What foods do you think they are just from looking at them?

    Carrots, apples, sugar, flour, etc.

3. Why do you think that?

    Because of the color or texture.

4. How can we test to see it that is what they really are?

    We can taste them, feel them, smell them, listen to the sound they make when we chew them.

5. Take a taste of one of your foods. Chew it slowly and carefully and try to identify it. Then take a taste of the other food and do the same.

6. What did you learn about your foods?

    They taste different. They feel different.

7. Can we always tell what a food is just by looking at it?

    No.

8. Were you able to identify both of the foods in your food pair?

    Yes/No.

9. What ways did you use to identify these two foods?

    I looked at the color, tasted them, felt them in my mouth, compared how one tasted or felt to the other.

10. If you couldn't always identify both foods, why couldn't you?

    (Answers will vary.)

Now let the students taste the foods in the other groups' food pairs. Advise students how to handle foods, toothpicks, and spoons in a safe and sanitary way (touching only the foods they will put on their plate, using a toothpick to pick up food pieces, using a spoon to pick up powered foods and put them on their plate, etc.) Tell students to place the foods from each pair together on their plate so they can compare them when they get back to their seats.

11. Take a taste of each food in a pair and compare them to each other.

12. What ways were the foods in pairs different from one another?

    One tasted good, the other tasted bad; one was juicier; one was sweeter; one was salty; one was hot and peppery; one was smoother/rougher/crunchier; etc.

— Practical Application —

Ask for a volunteer to try to identify something to eat that was not on the table. Give the following directions to the volunteer.

1. Hold your nose and breathe out of your mouth. Close your eyes. See if you can tell what I put in your mouth.

Take the peppermint out of your pocket, unwrap it, and put it into the student's mouth.

2. Keep holding your nose. You can open your eyes. What flavor do you taste?
    It doesn't have a flavor.

3. What do you think it is?
    A piece of candy.

4. Why do you think that's what it is?
    I heard you unwrap it. It's hard and shaped like a candy.

5. Stop holding your nose. Can you tell what flavor it is?
    Peppermint.

6. How do you know?
    (Students usually say they can taste it.)

7. Why do you think you couldn't tell what flavor it was when you held your nose?
    I couldn't smell it.

8. So did you need to be able to smell it to taste it?
    Yes.

9. We've found that there are quite a few ways we can use to identify foods. Let's see if we can name the ways.
    Taste them. Feel them. Listen to them. Look at them. Smell them.

Before distributing peppermints for the students to test, tell them not to open the peppermints until you give them directions. Distribute the peppermints.

10. Everyone unwrap your mint and hold it in your hand. Now hold your nose closed with the other hand and close your eyes. Put the peppermint in your mouth but keep holding your nose.

11. Wait a few seconds. Do you taste it? Now let go of your nose. Do you taste it now?

12. Now hold your nose again and wait a few seconds. Now let go of your nose.
13. Do you think smell is important for tasting?

     Yes.

14. Other words for smell are "odor" and "aroma." This candy has peppermint in it. There is a large family of plants that smell like peppermint or spearmint. They give off a strong smell into the air and are called aromatic herbs. We use flavorings from their leaves in foods, candies and gums, and medicines. Aromatic herbs have to be smelled as well as tasted to get their special flavor.

Take time to discuss how the students tasted the peppermints.

15. What foods that are left have a smell that helps you identify them?

     (Answers will vary.)

Students may want to test these foods by holding their nose for a taste test.

Have the students finish the foods if they would like to and safely dispose of their plates, spoons and utensils.

ACTIVITY 7: HOW DOES IT FEEL?

Goal: To experience that objects can be identified by feeling them

Skills: Observing by feeling, describing, identifying, classifying

Materials: Two large, clean socks for holding objects
Smooth objects to go in one sock (use objects that are familiar to most students):
- small, smooth ball
- plastic spoon
- piece of chalk
- rubber eraser
- marble
- crayon (with paper removed)

Objects with different textures, shapes, and sizes to go in the other sock:
- piece of sponge
- wash cloth
- drinking straw
- crayon
- small toy car
- piece of string
- whistle
- small square of sandpaper

Feel box – a covered box big enough to hold an assortment of objects—with a hole large enough for students to put their hands into but not be able to see the objects inside the box.

Preparation: Put the objects into the socks.

Preparation Time: 5 minutes

Lesson Time: 20–25 minutes

— Procedure and Questioning Strategy —

This activity can be done with small groups or the entire class. If you do the activity with the entire class, you may wish to fill the first sock with enough items for half the class to feel, proceed with the questioning strategy and identification, then fill the second sock and repeat the procedure.

1. This is a special sock with many different things in it. We are going to take turns reaching into the sock to feel one item. We do not want to tell anybody what it is we are feeling. We will only describe the object that we are feeling. For instance, we can tell things like whether the object is soft or hard, round or square, or things like that. When it is your turn, tell as much as you can about the object. The other students will ask you questions and try to guess what it is you are feeling.

Guide students to describe an object so it can be identified by other students. Ask prompting questions such as: Is it long or short, skinny or fat? Can you bend it or squeeze it? How big (or little) is it? What is it used for? How many parts does it have? These questions will also model the kinds of questions the other students should ask to obtain useful information.

After the class has guessed what an object is, have a student pull the object out of the sock so everyone can see it and feel it. Then tell the student to give the sock to someone else. Repeat the same procedure until all the objects have been identified.

2. When you felt an object in the sock, how did you know what it was?
   (Answers will vary.)

3. What did you use to explore and feel the object?
   My fingers. My hand.

4. What words did we use to describe the objects in the sock?
   (Answers will vary.)

5. Let's try to group some of those words. Which ones tell about shape?
   Long, short, skinny, fat, etc.

6. Which words tell how something feels?
   Soft, hard, etc.

Have the students continue grouping and subgrouping words they used to describe the objects. You may wish to write these words on the board or overhead. Now hold up the other sock.

7. We have another special sock with more objects in it. We'll take turns reaching into the sock. Remember to tell us as many things as you can about the object you are feeling. The rest of us will ask you questions and try to guess what the object is.

Have the students explore this sock just as they did the first one. Ask prompting questions such as: Is it furry or fuzzy, bumpy or flat? Does it have holes in it? How many parts does it have? Does it feel warm or cold? Can it open and close? Is it rough or smooth, thick or thin? Can something fit into it? Continue around the group until all students have had a turn or the sock is empty.

8. What different words did we use to describe the items in this sock?
   (Answers will vary.)

9. How can we group these words?

    By size, shape, texture, number or type of parts, temperature, etc.

Continue grouping the descriptive words. These words may be added to those written on the board or overhead. Ask the students if there are any ways to subgroup the groups of objects.

10. Now let's compare the objects within our groups. Which one is the roughest?

    (Answers will vary.)

Continue comparing the objects in the groups using words that students have selected for their descriptions.

— Practical Application —

Put the objects from the socks into a feel box. Leave the feel box on a table so students can continue to explore the objects. Add a couple of different objects each day.

* * * * *

When the students go out on the playground, take them on a science hunt. Ask them to help you collect small things from outside to put on a classroom science table. Each student could find one special thing that they like to feel. Objects like small twigs, bark, leaves, nuts, small rocks, small pine cones, and pine straw can be suggested. Make sure they know not to pick up objects that would be unsafe (like broken glass, sharp metal, etc.).

ACTIVITY 8: LOOKING CLOSER – WITH MAGNIFIERS

Goal: To explore familiar objects using magnifiers to see the objects differently

Skills: Observing, describing, comparing, explaining

Materials: As many different kinds of magnifiers as can be gathered, for example:
      inexpensive, unbreakable hand lenses
      box magnifiers
      jumbo magnifier on a tripod
      magnifiers of different strengths
    Objects found around the classroom like different types of paper,
      chalk, pencils, etc.
    Collections of objects. For example:
      rocks and sea shells
      soils
      feathers
      twigs and leaves
      dead insects
      seeds and grasses
      flowers

Preparation: Place magnifiers and objects where they are easily accessible to students.

Preparation Time: 5 minutes

Lesson Time: 25–30 minutes

— Procedure and Questioning Strategy —

Introduce students to magnifiers by showing them how to use each of the different types of magnifiers. Allow time for the students to explore on their own so they can see how objects appear when magnified (bringing the objects closer, moving them farther away, looking at them from the side, etc.).

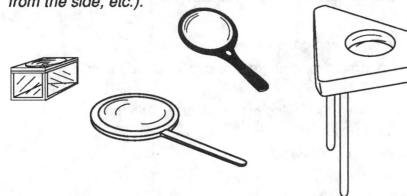

As students have experience with using magnifiers, ask them to describe what they see under the magnifier, then to compare the object as seen without a magnifier. Ask guiding questions like the following.

1. How does the object look different when you see it under the magnifier?

2. What do you see with the magnifier that you didn't notice before?

3. What does it look like when you see the object through a different magnifier?

— Practical Applications —

Suggest that students use magnifiers to look closely at different objects. For example:

> *shirt, pants, socks, and shoes*
> *back of hand, palm, fingernails*
> *objects in the room*
> *collections of objects suggested in "Materials"*

1. Do things always look different under a magnifier?

2. How do they look different?

3. Why do you think things look different?

Some students may notice that the lenses of magnifiers are thicker in the middle than around the sides. Let students know that the way the lens is curved makes it work as it does.

* * * * *

When students discover that magnifiers make things look larger, clearer, and more detailed under a magnifier, they usually want to examine closely as many things as possible. These are suggestions for further exploration:

> *newspaper or magazine print and pictures—including comics—in black and white and in color to notice dots printed close together*
>
> *skin – pores, lines, and hairs*
>
> *spider webs – intricate patterns and construction*
>
> *different foods – like bread, crackers, cheese, snacks, and foods brought for lunch*
>
> *different kinds of cloth, fur, velvet, stocking material, and satin to notice variety of weaves and textures*
>
> *wood – differences in texture and grain*
>
> *fingerprints – similarities and differences*
>
> *drawings and paintings – texture of pencil lines, crayon and paint colors and textures*
>
> *grass, sand, tree bark, pine needles, and leaves*

These kinds of items may be added to your science table for further investigation.

ACTIVITIES 1–8: CONNECTIONS

Goal: To find connections between the activities for observing with the senses

Skills: Observing, describing, comparing, classifying, categorizing, summarizing

Materials: Collections of objects from previous lesson

Preparation: Put together different sets of objects for students to classify.

Preparation Time: 15–20 minutes

— Procedure —

Divide students into cooperative groups. Give each group a set of objects from one of the previous activities. Encourage the students to look for similarities, then differences among the objects. Ask them how they could divide the objects into different sets or groups according to the similarities and differences they have found. Ask each group to explain why they grouped objects as they did.

Encourage the students to use different ways to classify objects using as many senses as possible. Remind them to look, listen, and smell.

Magnifiers could be used to see details in objects used for classification.

Ask the students to regroup the objects and to look for subgroups within their larger groups. Ask them what attributes they used to fit an object into a group. Ask them why they put an object in one group and not in another group. Share and discuss the variety of ways that objects have been grouped so students understand many different possibilities. Encourage precise, descriptive vocabulary.

Make pictorial charts and/or bar graphs of objects in different groups so students can see the grouping or hear how others think through their classifying process.

UNIT 2: WATER

ACTIVITY 9: EXPLORING WATER

Goal: To explore water
To observe that water flows when poured from container to container
To observe that different containers hold different amounts of water

Skills: Observing, classifying, inferring, predicting, discussing

Materials: Large tubs of water or a water table
Plastic pitchers (for filling containers)
Unbreakable containers such as:
plastic cups and containers of various sizes
assorted, unbreakable measuring cups
assorted sizes of tin cans with lids removed and sharp edges smoothed
Funnels (for filling small-mouthed containers)
Paper towels
Aprons (optional) – these can be made with a plastic bag using the instructions to the right.

Preparation:
1. Fill the tubs about two-thirds full of water.
2. If you are using the plastic aprons, help the students make them.
3. Set the tubs outdoors along with a supply of as many containers as possible. If the weather, or other conditions, make working outdoors difficult, put the tubs and containers on a table and spread newspapers on the floor of the classroom.
4. Help students put on their aprons.

Preparation Time: 10 minutes

Lesson Time: 25–30 minutes

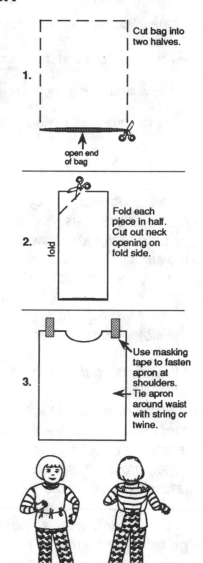

1. Cut bag into two halves.
open end of bag

2. Fold each piece in half. Cut out neck opening on fold side.
fold

3. Use masking tape to fasten apron at shoulders. Tie apron around waist with string or twine.

— Procedure and Questioning Strategy —

Select two containers of similar size. Use the pitcher to fill one of the containers about half full. Demonstrate pouring water from one container to another container, then back to the first container.

1. Describe the water as I pour it from one container to another.

   Possible responses: It runs into the other container. It bubbles. It splashes.

2. When water moves quickly from one container to the other, we say it runs in a stream or it flows. What other things flow when they are poured?

   Possible answers: juice, milk, syrup, sugar, salt, etc.

Select any container and use a pitcher to pour water into it so it is half full. Ask a student to point to the place on the container to show how high the water is in the container.

3. (Student's name) pointed to the water level in the container. When the water level is halfway up in the container, we say the container is half full.

Fill the container to the rim with water.

4. Where is the water level now?

   At the top of the container.

5. When the water level is at the top, we say the container is full.

Indicate all the assorted containers on the table.

6. These are all called containers. A container can hold something we put into it. Containers come in many different sizes. We're going to use these containers to hold water.

— Practical Application —

Students need to have many hands-on experiences to begin to understand conservation of volume. Some of these experiences can be included with students' exploration of water.

Allow students as much time as they can easily handle to explore with water. As you interact with the students while they explore, build on their comments about what they are doing by directing some activities and asking them questions.

1. Pour water back and forth from one container to another. Describe what the water looks like when you pour it.

   It's smooth. It twists around. There are bubbles at the bottom.

2. What do we say water does when we pour it?

   It runs in a stream. It flows.

3. When you pour all the water out of a container, what is left inside of the container?

   Small drops of water.

4. Why do you think the drops stayed in the container?

   (Answers will vary.)

5. Look at the different containers. Which are larger? Which are smaller?

6. Select a small container and use a pitcher to fill it half full with water. Point to the water level in the container.

7. Now add more water until the container is full. Now point to the water level.

8. What happened to the water level?

   It's higher. It's at the top of the container.

9. What do you think will happen to the water level if we pour the water in this container into a larger container?

   (Answers will vary.)

10. Try it.

11. Look at the water level in the larger container. Where is it? Is it as high as it was in the smaller container?

    No.

12. Why not?

    (Answers will vary.)

Guide students to compare the sizes of the containers' diameters.

13. What do you think will happen when you pour the water back into the smaller container?

14. Try it.

15. Where is the water level now? Why do you think the water level goes down in the big container and up in the small container? Is there more water in the smaller container?

    (Answers will vary.)

16. Fill the pitcher with water. How many of the smaller containers do you think you can fill with this much water?

    (Answers will vary.)

17. Try it.

18. How many containers did you fill?

    (Answers will vary according to sizes used.)

After the students have had enough time to explore water, discuss with them the things they have observed.

19. What did the water do as you poured it from one container to another?

    (Answers should include something similar to "the water flowed like a stream.")

20. Which container held the most water?

    (The biggest one.)

21. Which container held the least water?

    (The smallest one.)

22. When we poured the small container full of water into a larger container, what did the water level do?

    It went down.

23. When we poured it back into the small container, what did the water level do?

    It went back up.

24. Did the amount of water stay the same?

    Yes.

25. Did the size of the containers stay the same?

    No. The water went from a small one to a big one.

22. So what made the water level go up and down?

    The size of the container.

ACTIVITY 10: WATER – A CLOSER LOOK

Goal: To observe and describe some characteristics of water

Skills: Observing, classifying, describing, experimenting, explaining

Materials: Two plastic tumblers
Water
For each pair of students in the class:
two plastic eyedroppers
two pieces of waxed paper
two small paper cups – 3 or 5 oz.
paper towels

Preparation: 1. Set out all the materials on a table.
2. Pour water into one of the tumblers and into half of the paper cups.

Preparation Time: 3 minutes

Lesson Time: 25–30 minutes

— Procedure and Questioning Strategy —

Hold up the two tumblers.

1. Watch while I pour water from one tumbler into another one.

2. What do you notice about the water as it flows to the tumbler below?
   (Answers usually include that the water falls down.)

3. Does the water splash out in drops?
   No.

4. How does it fall?
   It falls in a stream.

5. Can you tell me some other places where you have seen water falling?
   (Possible responses: When you turn on the drinking fountain. Rain. Water running from a faucet. Waterfalls. Water in the shower.)

6. Which of these falls in a stream?
   (Possible answers: Water in the drinking fountain. Waterfalls. Water in the shower. Water from the faucet.)

7. Think about the rain. How does it fall?

   In drops.

Distribute pieces of waxed paper and eyedroppers to each student. Give each pair of students a paper cup with water in it.

8. Squeeze the bulb of the eyedropper as you put it into the water. See all the air bubbles. Now let go. What happened?

   Water went into the dropper.

9. Lift the dropper full of water above the water and squeeze the bulb very slowly. What happens this time?

   Drops come out. They fall into the water.

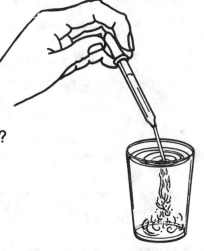

10. Fill your dropper again. Hold your dropper over the cup. Now quickly squeeze the bulb hard. What happens?

    The water comes out in a stream.

11. What did you do to make the stream?

    Squeezed the bulb hard and fast.

12. What did you do to the eyedropper bulb when you made drops come out?

    Squeezed it slowly.

13. How are drops of water different from a stream of water?

    The drops are smaller and rounded and come out separately. In the stream the water flows all together.

14. Drop some water drops at different places on your waxed paper. What do they look like?

    (They're round. Some are big, some are small.)

15. What does the top of each drop look like?

    Sort of rounded. Like a hill.

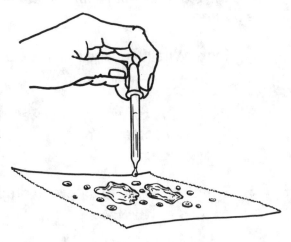

16. Try some experiments with your water drops on the waxed paper. Push some of the little drops together with your eyedropper. Then try to break up the big drop into little drops.

17. Try to make a drop as tiny as you can. Try to make a drop as big as you can.

18. Is there any difference in shape between the tiny drops and the big ones?

> The tiny drops are more rounded on top. The big drops spread out—they're flatter.

Give the students plenty of time to explore with the water. Remind them that they are experimenting to find out as much as they can about drops of water. Tell them that they will share what they have learned with the rest of the class when they are finished.

19. What happened when you put drops of water close enough together so that they touched?

> They moved together into a big drop.

20. What happened when lots of big drops flowed together?

> The water spread out. The water got flatter and made a puddle.

21. Where else do we see puddles?

> After it rains we see mud puddles or puddles in the road. When we spill water on the floor.

22. Did the rain puddles or mud puddles come down in one big drop?

> No. Little rain drops flowed together.

23. What shape are puddles compared to drops?

> Different shapes—all spread out, sort of round.

24. How did you make smaller drops from big drops?

> By separating them with the eyedropper. By breaking up the big drops.

— Practical Application —

Distribute an empty cup to each pair of students.

1. With your eyedroppers, squeeze some water in drops into the cup that has no water in it. Put as many drops in the cup as you can until I say, "Stop."

Give the students ample time to put enough drops into the cup so they have enough water to be poured.

2. Stop. What happened to the water drops you put into the cup?

> They all went (flowed) together.

3. Pick up the cup you just put water drops in and pour the water into your first cup. How did the water come out?

> In a stream.

4. What happened to the water drops?

> They stayed together.

5. Even when they were poured?
   Yes.

6. So, what's in a stream of water?
   Lots of water drops moving together.

* * * * *

Let students observe water flowing from the faucet and the water fountain. Have them observe and describe the similarities and differences between these observations and the way water looks when it is poured.

ACTIVITY 11: THE SHAPE OF WATER

Goal: To understand that water takes different shapes—It either takes the shape of its container or it spreads out on a surface

Skills: Observing, describing, comparing, predicting, explaining, testing predicting, generalizing

Materials: Transparent plastic tumbler
Pitcher (like a two-quart juice pitcher)
Water
Tall, skinny olive jar or graduated cylinder (50–100 ml.)
Transparent container that has a wide diameter like a large mayonnaise or pickle jar
Unbreakable containers of different sizes and shapes
Paper towels

Preparation: 1. Set all materials where students can see them.
2. Fill the pitcher with water.

Preparation Time: 2 minutes

Lesson Time: 10–15 minutes

— Procedure and Questioning Strategy —

Pour water from the pitcher into the tumbler and hold it up so all students can see it.

1. Describe what you see in my hand.

    A tumbler with water in it.

2. What shape is the water in the tumbler?

    The same shape as the tumbler.

3. Watch while I pour some of this water on the table.

Pour a small amount of water on a tabletop where all students can see the shape of water as it spreads out on the surface.

4. Describe the water on the table.

    It spread out. It's like a puddle of water.

5. How is the shape of the water on the table different from the shape it had in the tumbler?

    It's not as tall. It's wider. It's not as round. It's flatter.

6. Watch as I try to push the water into a pile. What happens?
    It spreads out again.

7. Why do you think the water spread out so far?
    There was nothing to stop it.

8. What stopped the water in the tumbler?
    The sides of the tumbler.

Pour water from the pitcher into the olive jar.

9. Describe the shape of the water in this jar.
    It's tall and skinny. It's rounded on the sides.

10. How is the shape of this water different from the shape of the
    water in the tumbler?
    It's taller and not so fat.

11. What made it this shape?
    The jar.

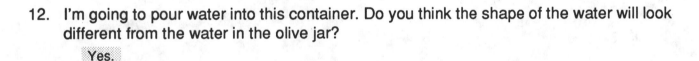

Hold up the wide-diameter mayonnaise jar.

12. I'm going to pour water into this container. Do you think the shape of the water will look
    different from the water in the olive jar?
    Yes.

13. How do you think it will be different?
    It will be wider (fatter).

14. Why do you think that?
    The container is wider.

Pour water from the pitcher into the container.

15. Good thinking! It is wider.

*Put the tumbler, jar, and container with water in
them where all students can see them.*

16. What is the same about what's in the tumbler, the tall, skinny jar, and the container?
    It's all water.

17. What's different about the water?

It's in different shapes.

18. Why do you think it has different shapes?

It's in different containers and the containers have different shapes.

19. What can we say about the shape of water?

It takes the shape of its container.

— Practical Application —

1. What other kinds of things take the shapes of their containers?

Possible answers: milk, juice, soda, cocoa, jello.

2. How do you get them into a glass or cup?

By pouring them.

Allow time for students to pour water between different-sized containers. Ask them to observe how the shape of water changes depending on the container it is in.

© 1991 *CRITICAL THINKING PRESS & SOFTWARE* • *P.O. Box 448, Pacific Grove, CA 93950*

ACTIVITY 12: SAME VOLUME, DIFFERENT SHAPE

Goal: To understand that the same amount of water can take different shapes

Skills: Observing, describing, comparing, classifying, explaining, predicting, testing, summarizing

Materials: Tall, skinny olive jar, or similar shaped container, or tall graduated cylinder (50–100 ml.)
Pitcher (like a two-quart juice pitcher) filled with water
Container with a wide mouth (like a quart mayonnaise jar)
Unbreakable containers of different sizes and shapes

Preparation: 1. Set out all materials where the students can see them.
2. Fill the pitcher with water.

Preparation Time: 2 minutes

Lesson Time: 10–15 minutes

— Procedure and Questioning Strategy —

Hold up the tall, skinny container.

1. Watch while I pour water into this container.

Use the pitcher of water to fill the container almost to the top.

2. Describe how much water is in the container.
   It's almost full.

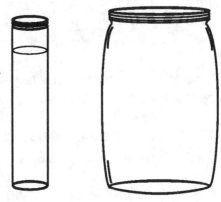

Hold up the wide-mouth container and the tall, skinny container filled with water.

3. How far do you think the water from the small container will go up in this bigger container?

Have different students point to predicted levels, marking each level on the container with an overhead projector pen.

4. Let's try it and see.

Pour all the water from the tall, skinny container into the wide-mouthed container.

5. What happened?
   The water didn't go up as far as we thought it would.

6. Why do you think that happened?
   The big jar is wider than the small jar.

If the students have trouble understanding this concept, wait until later to give them clues.

7. Let's pour the water from the wide container back into the tall, skinny container. Think to yourself about how far the water will rise in the skinny container.

Pour the water back into the tall, skinny container.

8. Did it rise to where you thought it would?

    Yes.

9. How far up did it go?

    As high as it was before it was poured into the wide container. Almost to the top.

10. Why do you think that happened?

    (Answers vary. Some students mention that no water was spilled and/or the same amount of water went back into the container as was in it to start with.)

11. Which container looked like it had more water in it?

    The tall, skinny container.

12. Why do you think it seemed that way?

    The water level was higher.

Empty the water out of the tall, skinny container. Hold the tall, skinny container and wide-mouth container so the students can see the tops of them.

13. Look at the tops of both of these containers. What's different about them?

    The tall, skinny container has a much smaller opening at the top than the wide container.

Show them the bottoms of the containers.

14. What's different about the bottoms?

    The bottom of the tall, skinny container is smaller than the bottom of the wide container.

15. Which container is bigger all the way around?

    The wide container.

16. Which is the smallest jar all the way around?

    The tall, skinny jar.

17. Do you remember what happened when we poured water on the table in the other activity?

    It spread out in a puddle.

Select a small cup.

18. Watch carefully while I fill this small cup full of water and pour it into the skinny container.

Pour a small amount into the skinny container.

19. What is the water level doing?

> It's going up the sides of the container.

Pour the rest into the container slowly as the students watch the water level rise.

20. I'm going to fill the cup up again and slowly pour it into the wide container.

Pour a small amount into the wide container.

21. What do you see the water doing?

> It's spreading out into a puddle.

Continue pouring water slowly into the wide container until it hits the sides and the water level begins to rise.

22. What happened?

> The water spread out until it reached the sides of the wide container, then it started to go up.

Finish pouring the rest of the water into the wide container.

23. How much water did I pour into each container?

> One cupful.

24. Which container let the water spread out more before it moved up the sides of the container?

> The wide one.

25. So, why doesn't the water level go up as far in the wide container as it does in the skinny one?

> It spreads out more.

26. So when water can spread out more in a wide container it does not go as far up the sides.

— Practical Application —

Have students pour water between different-sized containers so they can have more experience with the concept.

1. Experiment with the containers in front of you. Find which container lets the water spread out the most and which one lets it spread out the least.

ACTIVITY 13: EVAPORATION

Goal: To have experience with evaporation

Skills: Observing, describing, inferring

Materials: Several containers for water (for students to dip brushes and sponges)
Water
Large easel brushes or pieces of sponge
Two or three washcloths or handkerchiefs or pieces of similar weight cloth
Large pieces of aluminum foil
Paper towels

Preparation: Set out the materials on a table so all students can see them.

Preparation Time: 5 minutes

Lesson Time: 20–25 minutes

— Procedure and Questioning Strategy —

Let the students observe as you pour water into the containers. Set the containers in different places near the chalkboard. Put the brushes near the water containers. Then have the students come up to the chalkboard in small groups.

1. We're going to make water pictures. Think of something you could draw, dip your brush or sponge into the water, and make a picture on the chalkboard.

Have the students observe the pictures they have drawn.

2. What do you see happening?
   The picture of the (object) is disappearing.

3. What did we use to make the pictures?
   Water.

4. Where did the water go?
   Possible answers: It dried up. It went into the air. It soaked into the board.

Wait until the chalkboard has dried.

5. The chalkboard is dry now. The water went into the air. When water goes into the air, we say it *evaporates*.

6. Could we see the chalkboard getting drier?

> Yes.

7. Could we see the water going into the air?

> No.

Give all of the students a chance to draw a picture and watch it evaporate.

8. Did some parts of pictures evaporate faster than others?

> Yes.

9. Why do you think that happened?

> There wasn't as much water in those places.

10. What happened when there was more water in some places?

> It took longer to dry (evaporate).

— Practical Application —

1. I am going to wet some pieces of cloth and squeeze out most of the water. Next I will place these wet cloths on large pieces of aluminum foil. Some of these I will spread out flat. Then I will take other pieces of wet cloth and roll them into a ball. Now let's find a warm or sunny place to put them.

Have students help select a sunny or warm place in the room to put the cloths. If no such place is available in the room, select a sunny spot outside or elsewhere.

2. Which pieces of cloth do you think will dry first?

> (Answers vary.)

3. Why do you think that will happen?

> (Answers vary.)

4. We'll check them every hour or so to see what happens.

At intervals during the day, have students feel the cloths and describe how wet or dry they are.

5. Feel the tops of the flat cloths then lift them up and feel the bottoms. Which side feels drier?

> The top.

6. Now spread them out again.

7. Feel the cloths that are rolled into a ball. First feel the outside then unroll them and feel the inside. Where does it feel drier?

   On the outside.

8. Now roll these cloths up again and we'll check all of them again in a little while.

After the last observation:

9. Which cloths dried the quickest?

   The ones that were spread out.

10. Where did they dry first?

    On the top.

11. Which cloths dried the slowest?

    The ones that were rolled up.

12. Where did they dry first?

    On the outside of the ball.

13. Why do you think that happened?

    The water could evaporate easier.

14. Where did it dry the slowest?

    Inside the ball of cloth.

15. Why do you think that happened?

    The water couldn't get out as easily. It evaporated slower.

16. Where did the water go as the cloths dried?

    Into the air. It evaporated.

17. Could you see the water evaporate?

    No.

18. If you wanted something to dry quickly, what would you do?

    Spread it out. Put it in a warm or sunny place.

* * * * *

You can extend this activity another day by placing some cloths in the sun and in the shade. Use the questioning strategies outlined above so that students are able to understand the effect the sun has on evaporation. Also ask:

1. How does the sun make you feel?

   Warm.

2. How does the sun make the cloths feel?

   Warm.

3. How do the cloths in the shade feel?

   Cold or cool.

4. How does the sun help the cloths dry?

   It makes them warm. It helps the water evaporate faster.

5. How do we know this?

   We could feel how warm the cloths were in the sun. We know they dried faster than the ones in the shade.

ACTIVITY 14: CONDENSATION

Goal: To have experience with condensation

Skills: Observing, describing, inferring, predicting

Materials: Teakettle filled with water
Electric hot plate
Mirror large enough so students can see condensation on it and so it can be
held over hot steam safely
Pot holders or mitts

Preparation: Put the teakettle on the hot plate where all students can see it. Turn on
the hot plate.

Preparation Time: 5 minutes

Lesson Time: 10–15 minutes

— Procedure and Questioning Strategy —

1. There is water in the teakettle. I've turned on the hot plate so the water in the kettle will get hot. Listen carefully as the water heats.

2. What do you hear?
   Some noises.

3. Describe the noises. What do they sound like?
   (Rumbling, roaring, squeaking like a motor, etc.—answers will vary.)

4. Where do you think they're coming from?
   Inside the kettle.

5. What do you think is happening?
   The water is heating up.

Remind the students that the teakettle, the hot water inside the kettle, and the air around it is very hot. The kettle and the hot water inside should be handled very carefully.

6. Watch the kettle and the place just above the spout.

7. What do you see?
   (Varied answers. If students don't mention steam, tell them that it is steam.)

8. Where is the steam coming from?

    Inside the kettle.

9. How did we make the steam?

    By heating the water.

10. We are really looking at hot water evaporating or escaping into the air.

11. What do you think will happen when I hold this mirror over the spout?

    (Answers will vary. The students should mention that the mirror will get hot.)

12. That's true. That's why I'm going to hold it with the pot holder.

*Hold the mirror above the spout using the oven mitts.
Tell the students that steam can burn, so they should
not do this themselves.*

13. What do you see on the surface of the mirror?

    It's clouding (steaming) up. There are drops
    of water on the mirror.

14. Where did the drops come from?

    The steam.

15. Where does the steam come from?

    From the hot water in the kettle.

16. So what is steam made of?

    Drops of water.

17. We call the drops gathering on the mirror *condensation*. Let's see how big the drops of water are in steam. I'll dry off the mirror and let it cloud up again. Look at the surface of the mirror. What do you see?

    Lots of very tiny drops of water.

18. Let's collect more condensation. What do you see happening now?

    The drops are getting bigger. Some drops are running down the mirror and falling off.

19. What is happening to make the drops bigger?

    The little drops are flowing together.

20. Why are the drops running down the mirror and falling off?

    The drops are too big and heavy. They are falling in a stream.

— Practical Application —

1. Sometimes when you go outside early in the morning you see small drops of water, called dew, on the grass. Or if you walk through the grass in the morning, sometimes it makes your shoes wet. What happens to the dew after the sun comes out?

   It dries up. It goes into the air.

2. Why do you think that happens?

   The sun is warm and it heats up the drops of water. They evaporate.

3. Some days you may see the grass or a puddle steaming. What do you think is happening?

   Water is evaporating.

4. When you see the grass or puddle steaming, is the water evaporating slowly or quickly?

   Quickly.

5. How do you know?

   Because the steam we collected on the mirror had lots of tiny drops of water in it.

6. What about the cloths we put out to dry? Were they evaporating quickly?

   No.

7. How do you know?

   We couldn't see the steam.

ACTIVITIES 9–14: CONNECTIONS

Goal: To find relationships among the activities for water

Skills: Observing, predicting, describing, comparing, inferring, explaining

Materials: Two large glass jars, at least quart size, with lids (like wide-mouthed mayonnaise jars)
Potting soil
A couple of small plants suitable for a terrarium
Measuring cup
Water

— Procedure and Questioning Strategy —

1. Think about some of the things we have learned about water. What happens when something like water evaporates?

   It goes into the air.

2. Yes, and when water evaporates into the air, then gathers on a surface, like on the mirror we held over the kettle of boiling water, we say the water condenses.

3. Why do you think the bathroom mirror clouds up when you take a hot shower or a bath?

   The steam from the shower (bath) gathers (condenses) on the mirror.

4. Let's think about something else that may have happened to you. Imagine that you have just gone swimming. Your skin is all wet. What happens to the water on your skin if you don't use a towel to dry it off?

   It evaporates into the air.

5. Would that take a shorter or longer time than to dry off with a towel?

   A longer time.

6. Imagine that you did use a towel to dry off. Why would you hang the towel on a rack or out on a line to dry?

   So the water can evaporate from the towel.

7. What if you left the towel all balled up on the bathroom floor? Would it dry as fast as it would if it were hung up?

   No.

8. Why do you think it wouldn't?

   The rolled up cloths we tested didn't dry as fast as the cloths that were spread out.

Let's try an experiment to find out more about evaporation and condensation.

Put the two glass jars on a table where the students can see them. Have the students watch as you 1) pour a half cup of water into one of the glass jars and screw the lid on tightly, 2) put some potting soil into the other jar and plant the small plants, 3) carefully pour a half cup of water down the inner side of the jar, 4) screw the jar lid on tightly, 5) place both jars in a sunny window.

Have the students observe the jars visually and tactilely at intervals during the day. Ask students to keep a record of their observations at each interval. At the end of the first day, ask them the following questions.

9. **What do you observe?**

    (Possible observations: There are small drops of water on the inside of the jars. There are larger drops on the inside of the jar lids. The jars felt cold at first and got warmer during the day.)

10. **What do we call the drops of water on the glass and lids inside the jars?**

    Condensation.

11. **What do you think happened to make the condensation form?**

    The sun heated the water inside the jars.

12. **How did the drops get on the glass and under the jar lids?**

    It evaporated from the water at the bottom of the jars.

13. **How do you know it didn't come from outside of the jars?**

    You screwed the lids on tightly. Nothing could have gotten in from outside of the jars.

Observations during the next few days should be: Some of the drops are running/streaming down the glass inside the jar. Some of the drops are dropping ("raining") from the jar lid.

Ask the students to relate these observations to the activity with the teakettle and mirror. This would be a good time to relate the terrarium to the water cycle.

14. **What do you see happening in the jars that is similar to rain?**

    Water drops down from under the jar lids.

15. **Where do real rain drops come from?**

    Clouds.

16. **Where do the water drops fall in the first jar?**

    On the water.

17. **Where might real rain fall on the water?**

    Oceans, rivers, lakes, ponds.

18. Where do the water drops land in the second jar?

    On the soil and plants.

19. Does rain land on soil and plants outdoors?

    Yes.

20. We said that the sun heated the water in the jars so it evaporated. Then it condensed under the jar lids and dropped back down to the bottom of the jar. We said that the rain comes from clouds. Where do the clouds come from?

    Water that has evaporated from the soil, oceans, rivers, lakes, and ponds.

21. What makes this water evaporate?

    The sun heats it.

22. We have just looked at the parts of the water cycle. Let's review these parts of the water cycle as I draw them on the board.

If you prefer, reproduce and hand out the drawings of the water cycle on the next page.

After you have reviewed the water cycle, have the students draw their own pictures of the water cycle or have groups each design a part of the cycle which they can then put together as a mural.

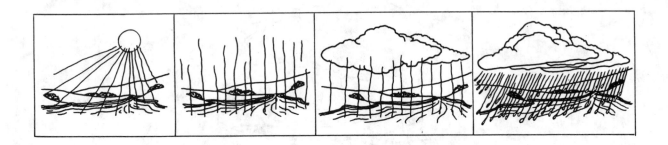

1. The sun heats the water on the earth.

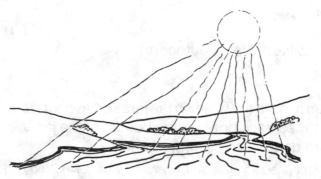

2. Some of the water evaporates. The water that evaporates rises up into the air.

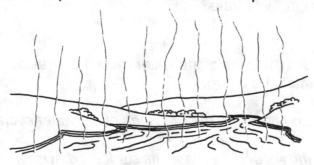

3. It condenses up in the sky, making clouds.

4. Then the water falls back to earth as rain.

UNIT 3: BUOYANCY AND SURFACE TENSION

ACTIVITY 15: CAN IT SINK? CAN IT FLOAT?

Goal: To explore the ways that objects sink and float

Skills: Observing, comparing, classifying, categorizing

Materials: 1 set of the following items for each small group
of students:
large container of water
a set of objects to test that are the same for each
group of students. For example:
wooden popsicle sticks
plastic knives, spoons, and forks
aluminum foil
pieces of sponge
styrofoam
coins
bottle caps
clothespins
leaves, twigs, pebbles, small rocks, acorns—
all gathered from outdoors
crayons
different kinds of paper, including tissue paper
and waxed paper
pencils
other objects found in the classroom and at home

Preparation: 1. Place containers of water and sets of objects to
test on tables around the room.
2. Prepare a sheet of chartpaper to list students'
responses (like the one shown to the right).

Preparation Time: 10 minutes

Lesson Time: 25–30 minutes

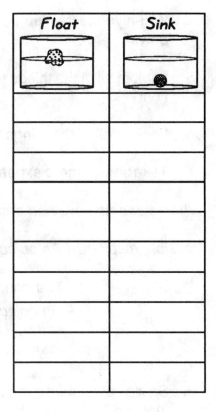

Float	Sink

— Procedure and Questioning Strategy —

Tell the students that they will work in groups to find out which objects sink and which float. First, they are to decide, as a group, if they think an object will sink or float. Then they will test the object.

Divide the students into groups, one group for each set of materials. Allow ample time for the students to test all of their objects, then discuss, as a class, which items sank and which floated. As the students discuss whether each item sank or

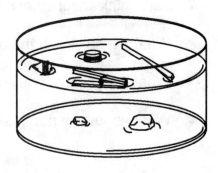

floated, list each object by name or quickly sketch it in the appropriate column on the chartpaper. Help students continue their discussion by asking guiding questions.

1. Look at all of the things that float. What is similar about them?

2. Look at all of the things that sink. What is similar about them?

3. What is different about the things that float and the things that sink?

4. When you test an object, see if it matters whether you turn it on its side or turn it upside down?

5. Look at all of the things you found that sank. Can you think of a way to make any of them float?

6. Look at all of the things that floated. Can you think of a way to make any of them sink?

7. Take some time to experiment and test your ideas, then we'll share what we have learned.

When students have had enough time, discuss the results.

Possible results: – *An object may float if they place it gently on the water, if they lay it horizontally (flat) on top of the water, if they place it on top of an object that can float, or if they flatten it out.*

– An object that usually floats may sink if they put it vertically into the water, if they change its shape to make it more compact, or if they leave it in the water until it gets waterlogged.

8. So we have found a group of items that float easily, a group that sinks easily, and a group that can float or sink depending on how we place them in the water or change their shape.

— Practical Application —

Answers will vary for the following questions.

1. Let's think about other things that sink or float. What kinds of things float in your bathtub?

2. What things sink in the bathtub?

3. What things have you seen that float or sink in a puddle or pond?

4. What about in a pool, lake, river, or stream?

ACTIVITY 16: ROUNDED WATER?

Goal: To understand that the surface of water can have a rounded shape

Skills: Observing, describing, comparing, explaining

Materials: Gallon container (like a plastic gallon milk container or large pitcher) filled with water
For each pair of students:
1 small medicine measuring cup (about 30 ml.) from a drug store, or a tiny, pleated paper cup from a restaurant supply store, or a 35mm film container
1 small paper or plastic cup (about 5 oz.)
10 pennies
2 paper towels

Preparation: 1. Set the materials out on a table.
2. Divide students into pairs.
3. Have one student from each pair get the materials listed above.

Preparation Time: 5 minutes

Lesson Time: 15–20 minutes

— Procedure and Questioning Strategy —

1. Each group should have two cups—a very small cup and a larger cup. I am going to come around and pour water into each group's large cup. Please wait until I have poured water for everyone. Then we will try something fun with the water.

Pour water into each pair's larger cup.

2. Each cup has a brim. The brim is here (*point to the brim*) at the top edge of the cup. Watch while I fill a small cup to the brim using the water from the larger cup. The water should come just to the top of the cup.

Demonstrate how to fill the small cup just to the brim using the larger cup.

3. Now I want one person in each pair to very carefully pour water into your smaller cup until the water just comes to the brim of the cup.

Check the water levels of the students small cups to make certain the water is just to the brim and not brimming over. There should not be a dome on the water. If there is, have students pour some of the water back into their larger cup.

4. How many pennies do you think could fit into the water in this small cup before the water spills over?

   (Answers vary. Students usually say that one might fit in, or none.)

5. I want the other person in each pair to see how many pennies he or she can put into the small cup of water without spilling any water over the sides of the cup.

   (8–10 pennies usually fit.)

Have the students share how many pennies they got into the water and how they did it.

6. Let's try it again! Pour the water from your small cup back into your larger cup. Dry the small cup and pennies with a paper towel. This time see if the other person in your pair can get more pennies into the cup than before. See how the water looks as you add each penny. Look from the side of the cup and watch what happens to the water as the pennies are added.

Give the students time to complete this new experiment.

7. Did you get more pennies in the cup this time?

   (Answers will vary.)

8. How did you do it? What way (technique) did you use that your partner didn't use the first time?

   (Answers will vary.)

9. What did the water look like as you put a penny into it?

   It sort of dented under the penny.

10. What happened after that?

    The penny broke through the water and fell to the bottom.

11. What happened to the water as you added more and more pennies?

    It got more rounded on the top.

— **Practical Application** —

1. Let's see if we can find out why the water gets rounded.

Have about eight students stand in a straight line with their shoulders almost touching.

2. You're going to be the top of the water in the glass. Imagine that I am going to add pennies to your water. Hold hands. The people on the ends of the line stay where you are. As I add each penny, the rest of you step slowly backward until I say, "Stop." "One . . . two . . . three . . . four . . . ."

When the line of students has become rounded and students cannot move farther backward without breaking their hold, stop adding pennies.

3. "Stop!" Take a look at your line. What does it look like?

   It's curved.

4. How is the shape of your line like the shape of the water on top of the glass?

   They're both rounded.

5. So what do you think the water on top of the glass was doing as you added pennies?

   Holding together.

6. When was the water holding together the hardest to try to keep from breaking?

   Just before it spilled over the side.

7. Why do you think it finally spilled?

   The water couldn't hold together anymore.

8. Did it let go all over the place at once?

   No. Just in one place.

ACTIVITY 17: FLOAT A PAPER CLIP!

Goal: To understand that the surface of water can withstand some pressure before it breaks

Skills: Observing, comparing, describing, inferring

Materials: Gallon container (like a plastic gallon milk container or large pitcher) filled with water
Transparent plastic cup
For each pair of students:
1 paper or plastic cup
4 standard size metal paper clips
paper towel
magnifying glass

Preparation: 1. Set the students' materials where they can reach them.
2. Divide the students into pairs.

Preparation Time: 5 minutes

Lesson Time: 15–20 minutes

— Procedure and Questioning Strategy —

Have one person from each pair get the materials for this experiment. Pour water to just below the top of each pair's cup. Fill a transparent plastic cup with water to just below the top of the cup. Then demonstrate how to do the first three steps of the experiment. Tell the students:

1. I'm going to unbend one paper clip . . .

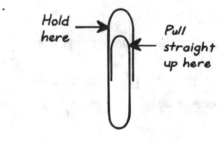
Hold here →
Pull straight up here

. . . so it looks like this.

2. Then I'm going to lay another paper clip across the bottom of the first clip like this.

3. Now I'll use the clip I've unbent to gently lower the other clip on to the top (surface) of the water like this. Then I'll slowly move the unbent clip down a little and slide it out from under the other clip so it can float on top of the water.

4. Try doing this in your pairs. Remember to lower your clips slowly and gently into the water. Once you have the clip floating on the water, use your magnifying glass to look at the water around the clip.

Allow the students enough time to master the technique of floating the paper clip. Then ask them the following questions.

5. What does the water around the paper clip look like?

    It looks like it's attached to the clip. The clip looks like it's making a dent in the water.

6. Take another clip and hold it by one end above the water. Let it drop into the water.

7. What happened?

    It sank.

8. Why do you think it sank this time instead of floating?

    It was just dropped into the water. The end of the clip is smaller than the long part of the clip we floated.

9. Take another clip and slowly and gently push one end of it into the water until it breaks through the water.

10. What did the top of the water look like?

    It looked like it was dented until the clip broke through it.

11. Why do you think the clip broke through?

    It pushed through. I pushed it through.

— Practical Application —

1. How do you hold your body when you float on the water?

    Out flat—lying down.

2. When you float, how is your body like the floating paper clip?

    They both float when they lay flat on the surface of the water.

3. Could you float if you were in an upright, or standing, position?

    No, not unless I was wearing a float or swim vest.

4. If you were going to dive into the water, would you want to hit the water in a flat position or head first?

    Head first.

5. Why would you want to dive in head first?

    It would hurt to do a belly flop.

6. Would you go into the water easier head first?

    Yes.

7. How is diving head first into the water like holding the paper clip at its end and dropping it into the water?

    Both go through the water with the smaller end first.

ACTIVITY 18: HOW DO BOATS FLOAT?

Goal: To recognize that the shape of something can determine whether it sinks or floats
To find a way to make something that usually sinks be able to float

Skills: Observing, classifying, designing, experimenting, drawing conclusions

Materials: Water
For each group of students:
container that will allow a large surface area of water
modeling clay—half a stick
paper towels
2 squares of aluminum foil—about 10 cm. (4") square
8–10 marbles

Preparation: Fill the containers with water and place them on tables around the room, one container per group.

Preparation Time: 5 minutes

Lesson Time: 25–30 minutes

— Procedure and Questioning Strategy —

Divide the students into groups. Give each group half a stick of clay and a paper towel.

1. What do you think would happen if you rolled your clay into a ball and dropped it into the water?

    (Students usually say that it would sink.)

2. Have one member of your group try it.

    (The clay ball sinks.)

3. Why do you think it sank?

    It's heavy.

4. Dry your clay with the paper towel. Now see if you can do something with your clay so it would float. Then test it in the water.

    (Most students flatten out the clay or make some object that will not float.)

Give the students time to test their clay designs, then to redesign them and test again. If they don't design a shape that floats, give them more clues. For example:

5. What shape did you make your clay?

    (Answers will vary.)

6. Can you think of anything that is made of heavy material that floats in the ocean or on a lake?

   A boat or a ship or a canoe.

7. Try designing something like that.

Each group should design a boat-like vessel. After each group designs a shape that floats, discuss their designs.

8. What did you do to make your clay float?

   Made it in the shape of a boat.

9. Let's compare and contrast our boats. Which boats seem to float higher in the water?

   The boats with thinner sides and bottoms. The boats that are longer or wider.

10. Carefully dry off your clay boats and put them on the table. We'll use them again later.

— Practical Application —

Give each group a square of aluminum foil and a marble.

1. What do you think would happen if you dropped a marble into the water?

   It would sink.

2. Try it.

3. What do you think the aluminum foil would do if you dropped it into the water just like it is?

   It would float.

4. Try it.

5. See if you can make something with the foil that will help the marble float. Wait to test yours in the water until all the groups have finished making theirs.

   (Students usually curve the aluminum foil around the marble.)

6. Okay, let's test them.

   (The marble and foil usually sink during the first trial.)

Give the students another piece of foil and a chance to design another shape that will help the marble float. Have them test this new shape. If they have not made their foil into a boat shape, tell them to think about what they did with the clay. Discuss with them the designs that seem to float the marble the best. Compare and contrast those that floated with those that sank.

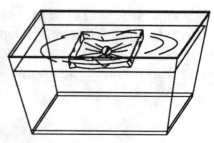

7. Now, put your clay boat back in the water. See what happens if you put a marble in the boat.

   It sinks lower in the water.

8. Add more marbles. See how many your boat can hold without sinking.

9. What happened to the boat as the marbles were added?

   It floated lower and lower in the water.

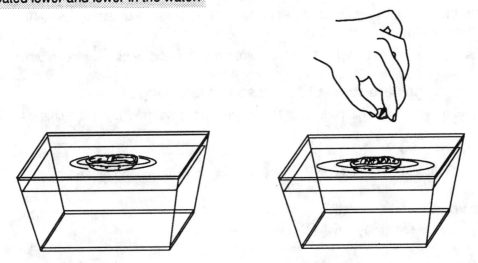

10. Many boats carry people and supplies like food, cars, and oil. What do you think happens as more people and supplies are loaded onto a boat?

    It floats lower in the water.

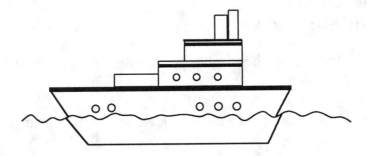

ACTIVITIES 15–18: CONNECTIONS

Goal: To find relationships among the activities for buoyancy and surface tension

Skills: Comparing, contrasting, classifying, discussing, explaining

— Procedure and Questioning Strategy —

1. We have put lots of things into water to see if they floated or sank. Let's name some of the things that floated.

   (Possible answers: leaves, bottle caps, aluminum foil, and clay boats, among others).

2. What do you think are some of the reasons these things floated?

   They were light in weight. They were spread out on the water. They had a shape like a boat.

3. Which things sank?

   (Possible answers: marbles, balls of clay, stones, crayons, among others).

4. Why do you think they sank?

   They were kind of heavy. They had the wrong shape.

5. What did we do to make some things that sank be able to float?

   We put them on something that floated. We flattened them out or changed their shape to make them look like boats.

6. Remember how many pennies we could add to water without spilling any water? Why didn't the water spill out?

   It was holding together on the top.

7. What shape was the water on top after you added some pennies?

   Rounded.

8. We found that paper clips usually sank, but we could float one on the water. Why did it float?

   We carefully placed the clip flat on the surface of the water.

9. How did the water look around the clip?

   The surface of the water dented a little and came up around the clip.

10. When water holds together on the surface, it makes it easier for something to float on it. The water seems to have a "skin" on the surface that helps hold things up. It helps things float on top.

11. What happened when you slowly pushed the end of a paper clip into the water?

The clip looked like it made a dent in the top of the water before it broke through.

12. So, we said that things that float are usually light in weight or spread out on the water. What else helps them float?

The way the water holds together on the surface—the "skin" on the surface of the water.

UNIT 4: AIR

ACTIVITY 19: EXPLORING AIR

Goal: To explore some of the properties of air
To understand that air can be felt when it can't be seen and that air takes up space

Skills: Observing, describing, comparing, classifying, explaining, generalizing, supporting inferences with observations

Materials: One for each student:
paper bag (lunch size)
twist tie
balloon
Felt-tip markers
String or yarn
Scissors
Party horns and/or whistles

Preparation: 1. Set the materials on a table where students can reach them.
2. Distribute paper bags and twist ties so every student has one of each.

Preparation Time: 5 minutes

Lesson Time: 20–30 minutes

— **Procedure and Questioning Strategy** —

Demonstrate how to flatten out a paper bag.

1. Flatten the paper bag out on the table as I am doing. Now open it up again and look inside. What's in the bag?

   (Students usually answer, "Nothing.")

2. Flatten the bag again. What do you feel as you flatten it?

   Like I'm pushing something out of the bag.

3. Pretend your bag is a balloon. Do something to make your bag look like a balloon.

4. What did you do?

   I blew into the bag. I blew up the bag.

5. What did you blow into the bag?

   Air. My breath.

6. Blow as much air as you can into the bag and close it at the top. Put the twist tie tightly around the top of the bag to keep the air in the bag. What does the bag feel like now?

It feels full (or puffy or hard), like it has something in it.

7. What's taking up the space inside the bag?

    Air.

8. How do you know?

    We blew air into the bag. That's all that's in it.

9. Take off the twist tie. Squeeze the air out of the bag—fast! What happened?

    The bag flattened out. The air went out. It made a noise.

Distribute balloons, one for each student.

10. Hold both ends of your balloon and stretch it out a few times. Then blow up the balloon. Hold the balloon closed so the air won't escape.

11. What's in the balloon?

    Air.

12. How do you know that?

    I blew it in. The balloon is bigger. It feels fat.

13. What is taking up the space inside the balloon?

    Air.

14. Hold the opening of the balloon near your neck. Let the air out of the balloon. What do you feel?

    Air hitting my neck.

15. What did you hear?

    A noise when the air came out.

16. When we first looked into the paper bag, we thought there was nothing in it. But there really was something in it. What was it?

    Air.

17. How do we know that air was in the bag and the balloon?

    We blew in it. We felt it. We heard it go out.

— **Practical Application** —

1. Flatten out the balloon on the table. Push all the air out of it. Draw a picture or face on the balloon with a felt-tip marker.

2. Blow up the balloon again, little by little, and keep looking at your drawing.

3. What happened?

   It got bigger and bigger.

4. Why do you think that happened?

   I put more and more air into the balloon. The balloon stretched.

5. When did the balloon take up the most space?

   When it had the most air inside it.

6. So what happens as you put more and more air into your balloon?

   It gets bigger. It takes up more space.

7. Blow up your balloon again so it is big.

8. Now, make the opening of the balloon different sizes as you let the air out of the balloon.

Allow students enough time to blow up the balloon several times and experiment with different sized openings.

9. What do you notice about the sounds the air makes as it leaves the balloon?

   Some are higher (or lower) than others.

10. Blow up your balloon again and let it go. What happened?

    The balloon flew around the room. It made a funny noise.

11. Why did it do that?

    The air came out of the balloon really fast. It pushed the balloon around the room.

Have the students blow up their balloons and hold them so no air escapes. Tie each balloon securely at the neck with a piece of string. Leave the balloons overnight on a table. The next day discuss the changes in the balloons and the possible causes, such as "the air got out of the balloon" and "the string wasn't tight enough."

* * * * *

Blow through some party horns and whistles. Discuss what causes the horn's noises and what makes the whistles unfold and inflate.

ACTIVITY 20: AIR CAN BE ANYWHERE

Goal: To understand that air takes up space and can be found anywhere

Skills: Observing, describing, comparing, explaining, summarizing, supporting inferences with observations

Materials: Package of lunch bags
Items found around the classroom (like blocks, boxes of crayons, books) that can fit inside a lunch bag

Preparation: 1. Do this activity with small groups of students, (4–5 students at a time) or as a whole class activity with students sitting in a semi-circle on the floor.
2. Set the materials close to where they will be used. Make sure the bags are as flat as possible.

Preparation Time: 5 minutes

Lesson Time: 10–15 minutes

— Procedure and Questioning Strategy —

Hold up the paper bags.

1. Tell me something about these bags.
     They're flat. There's nothing in them.

2. How do we usually use paper bags?
     We put things in them. We put our lunch in them.

Hand a bag to each student.

3. Now I will hand you an object and you put it into your bag.

After you have given each student an object, show the students a flattened, empty bag.

4. This is how your bag looked before you put something in it. How has your bag changed?
     There's something in it. It's fatter. It's lumpier. It's heavier.

5. How do you know something is in it?
     You can see it. You can feel it.

Have the students watch as you fill a flattened, empty bag with air by holding one side of the opening and pulling the bag quickly through the air so the air pops it open. Then hold the bag tightly closed so the air doesn't escape.

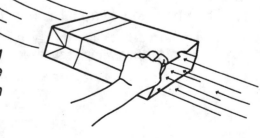

6. How did this bag change?

   It's fatter. It was flat then it opened up.

Let the students feel and gently squeeze the bag.

7. Is there something in the bag?

   Yes.

8. How do you know?

   It feels hard. It feels puffy. I couldn't flatten it very easily.

9. What is in the bag?

   Air.

10. Can you see it?

    No.

11. How did it get there?

    It went in when you pulled the bag through the air.

— Practical Application —

Have the students take the objects out of their bags. Now let them capture air from different parts of the room and outdoors too (when possible). Guide them to describe what they have in their bags and how they know what is there.

1. What did we find in other parts of the room and outdoors?

   Air.

2. How do you know that?

   We caught it in our bags.

3. How do you know that it was air?

   It filled up the bags.

4. Could you see the air in your bags?

   No.

5. Then how could you tell it was there?

   I could feel it. It made my bag open and it felt puffy when I closed it. I could feel air coming out of the bag when I squeezed it.

ACTIVITY 21: AIR DOES TAKE UP SPACE!

Goal: To understand that when air is contained, the amount of space it takes up cannot be taken up by something else

Skills: Observing, describing, predicting, explaining, summarizing

Materials: Large transparent container
Water
Empty soda bottle, without cap
Paper towels
Balloons
Large funnel
Masking tape
Container of water (like a plastic pitcher)

Preparation: Place the materials where all students can see and reach them.

Preparation Time: 5 minutes

Lesson Time: 15–20 minutes

— Procedure and Questioning Strategy —

Fill the large, transparent container about 2/3 full of water. Now hold up the soda bottle.

1. Describe this bottle for me.
   (Descriptions vary.)

2. What's taking up the space inside the bottle?
   Air.

3. What do you think will happen when I put this bottle, upside down, into the water?
   Water will get into the bottle. The bottle will get wet.

Turn the bottle upside down and plunge the bottle, top first, straight down, into the large container of water. Ask a volunteer to describe what s/he sees, especially where s/he sees the water level in the bottle. (The water level will depend on the size of the bottle opening and on how much water is in the container.)

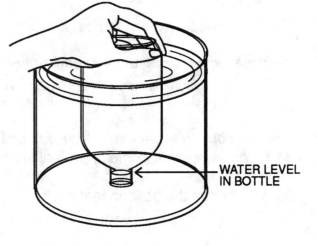
WATER LEVEL IN BOTTLE

4. Why didn't the water rise up very far in the bottle?
   The bottle opening is small. There was air inside the bottle.

5. Why do you think the air didn't come out of the bottle and let the water in?

    The bottle opening is small. The water held the air in the bottle. The air couldn't get out through the water.

6. Let's try something else.

Have the students observe as you blow up a balloon. Hold the neck of the balloon so no air escapes. Ask a volunteer to help you put a funnel into the opening of the balloon and tape the balloon neck tightly to the funnel so no air can get out. Keep holding the neck of the balloon below the funnel so no air escapes.

7. Describe what we have here.

    The balloon is blown up. The balloon is taped to the funnel. You are holding the balloon below the funnel.

8. What is taking up the space inside the balloon?

    Air.

9. What do you think will happen if we fill the funnel with water and I let the balloon go at the neck?

    (Predictions vary.)

Hold the balloon and funnel over the large water container. Ask for one volunteer to hold his or her hands under the balloon to support it. Ask another volunteer to fill the funnel with water. Tell everyone to watch closely as you release your fingers to allow the water in the funnel to flow into the balloon.

10. What happened?

    The air bubbled up in the funnel and made the water splash out. Then the water slowly went down into the balloon.

11. Why do you think air bubbled up through the funnel of water?

    Air was coming out of the balloon into the funnel of water.

12. What was making the air come out of the balloon?

    The water in the funnel was going down into the balloon. It was pushing the air out of the balloon.

Continue pouring water into the balloon until it is almost the same size as when it had air in it. Ask for a volunteer to come up to feel the balloon.

13. What does the balloon feel like?

    Like it has lots of water inside it. Like a water balloon.

14. Can you see where the water is in the balloon?

     Yes.

15. How high does the water go up into the balloon? Point to the water level.

     The water is almost up to the top of the balloon.

16. Can you see any air in the balloon?

     No.

17. If there were air in the balloon, where might it be?

     At the very top. Above the water level.

18. We said that the air was pushed out of the balloon by the water. How do you think the water could do that?

     It's heavier than the air.

19. How do you know that?

     A balloon full of air is light, it floats. A balloon full of water is heavy and has to be held up from the bottom.

20. So, what can we say about air and the space it takes up?

     It takes up space until something pushes it out of the space.

ACTIVITY 22: WHAT'S IN AN "EMPTY" CUP?

Goal: To understand that air takes up the space in what appears to be an empty cup

Skills: Observing, describing, predicting, generalizing, supporting inferences with observations

Materials: 7 plastic containers (transparent, if possible), large enough and deep enough
to fit a plastic tumbler inside
7 transparent plastic tumblers
Roll of paper towels
Pitcher of water (gallon size)

Preparation: 1. Divide the students into 6 groups.
2. Distribute to each group:
1 plastic container
1 plastic tumbler
3 or more paper towels
3. Put enough water into each group's container so that they can completely
submerge their tumbler in it.
4. Place the demonstration set of materials where all students can see them.

Preparation Time: 5 minutes

Lesson Time: 15–20 minutes

— Procedure and Questioning Strategy —

1. I am going to crumple a paper towel and squash it into the bottom of this tumbler so it won't fall out when I hold the tumbler upside down.

Hold the tumbler upside down so the students can see that the paper towel stays in the tumbler.

2. If I put this tumbler that has the paper towel in it upside down into this container of water and hold it on the bottom of the container, what do you think will happen?

   The water will go into the tumbler. The paper towel will get wet.

3. I'm going to let each of you find out what happens. Take turns trying this in your groups:
   1) Squash a paper towel into the bottom of your tumbler and turn the tumbler upside down.
   2) Put the tumbler straight down into the container of water and hold it on the bottom.
   3) Observe how it feels to hold the cup down in the water.
   4) See how high the water rises up in the tumbler. (Students may need to jiggle the tumbler slightly to see the water level.)
   5) Observe what happens to the paper towel.
   6) Discuss in your group what you think is happening.

Walk around among the groups to make sure each student has a chance to try the experiment and to make observations.

4. What did it feel like when you held the tumbler upside down in the water?

   > It was kind of hard to hold down. It felt like something was pushing it back up.

5. What happened to the paper towel?

   > Nothing. It didn't get wet. It stayed in the tumbler.

6. Why do you think it didn't get wet?

   > The water didn't go up that far in the tumbler.

7. How far did the water go up?

   > Just a little bit.

8. What was in the tumbler already that kept the water from going very far into the tumbler?

   > Air.

9. What happened to the air when you pushed the tumbler into the water?

   > The water pushed the air up into the tumbler.

10. What kept the paper towel from getting wet?

    > The air was all around the towel so the water couldn't get to it. The water was held back by the air.

11. If anyone has a paper towel that got wet, what did you do that let the water reach the towel?

    > I didn't put the cup straight down so the air got out and the water got in.

— Practical Application —

1. For centuries, people used simple diving bells to do a variety of tasks underwater. These bells were usually shaped like a large cup or bucket that was turned upside down and put into the water.

2. Can you think of a way that this sort of container could help people breathe underwater?

   > People could swim up under the bell and breathe the air that was trapped inside.

3. Yes, using a simple diving bell, people could work underwater for short periods, swimming into the bell to breathe until the air in the bell ran out. Then they would have to stop working or bring the bell back to the surface of the water to get more air.

4. In more modern times, the diving bell often had seats inside and was attached by a chain to a ship that floated on the surface of the water. The ship pumped a constant supply of fresh air through a hose into the bell. People could work a lot longer underwater with this kind of diving bell.

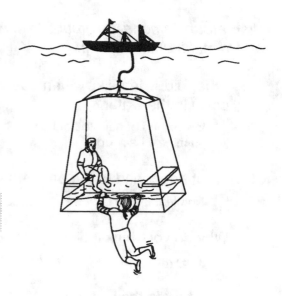

5. Why do you think it was important to keep pumping air into this type of diving bell?

   The workers would use up all the air unless more was pumped in all the time.

6. Today, diving bells are like underwater rooms. Small ones are attached to ships or dock rigging where they can be lowered or raised as needed. Divers are able to swim inside and rest, get a full tank of air, and make repairs to equipment. There are large ones that are self-contained, which means they don't have to be attached by a hose to a ship or go back to the surface in order to get fresh air. They are mounted on stilts on the sea bottom. They have tanks of pressurized air, food, beds so divers can rest, and even telephones and television. This way, divers can stay underwater for days or weeks at a time working at their tasks and living comfortably.

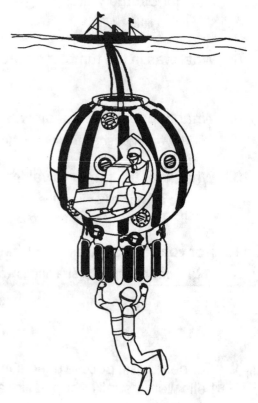

7. How do you think this last type of diving bell keeps a supply of fresh air without being attached by a hose to a ship above the surface of the water?

   Air is provided from air pressure tanks attached to the diving bell.

© 1991 CRITICAL THINKING PRESS & SOFTWARE • P.O. Box 448, Pacific Grove, CA 93950

ACTIVITY 23: WHAT'S IN A BUBBLE?

Goal: To understand that air is in bubbles

Skills: Observing, comparing, describing, classifying, explaining, predicting

Materials: Bubble solution in a large container
Several containers—1- or 2-quart size—or pans used for baking bread or cake
1 paper cup for each student
2–3 plastic straws for each student
Newspapers

Preparation:
1. Prepare the bubble solution. (The solution works best if it's made the day before it's used.)
2. Divide the solution into smaller containers.
3. Make a small hole with a pencil point about an inch from the bottom of each paper cup. The hole should be slightly smaller than the straw so the straw fits snugly.
4. Spread newspapers where students will be working with the bubble solution.
5. If the weather permits, you might find it best to do this activity outdoors.

Preparation Time: 10 minutes

Lesson Time: 20–30 minutes

> ### Recipe for Bubble Solution
>
> 10 cups of cool water
> 1 cup of dishwashing detergent
> (Joy or Dawn works best)
> 1 tbsp. liquid glycerine
> (from a drug store)
>
> Slowly stir to mix the solution, creating as little foam as possible.

— Procedure and Questioning Strategy —

Demonstrate the directions you give the students.

1. Carefully push your straw so it is about halfway into the hole in your cup.

2. Put the cup, upside down, into the bubble solution, then lift the cup up and look inside.

Show the students the opening of the cup and the bubble film inside it.

3. Describe what you see in the cup.
   There's a bubble in the cup.

4. Yes. It's a film of bubble solution. After you see the film, hold your cup upside down and slowly and gently blow through the straw.

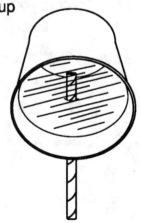

5. Let's try it. Remember to carefully push your straw through the hole in the cup.

Distribute one paper cup and one straw to each student. Place the smaller containers of bubble solution where the students can easily reach them. Tell them not to let the bubble solution get into their mouths because it tastes bad. Tell them to make certain they always blow out through their straws. As the students are exploring with the bubbles, ask them guiding questions.

6. What happened when you blew your bubbles with the cup upside down? What happened when you blew your bubbles right side up? Which bubbles were bigger?

    (Answers will vary.)

7. Try to make the biggest bubble you can.

8. Now try to make a small bubble.

9. What is inside your bubbles?

    Air.

10. How did it get there?

    We blew it in.

Let the students freely experiment as long as most students show interest. Then gather them together to discuss what they found.

11. When did you get the largest bubbles—when you held the cup upside down or right side up?

    Upside down.

12. How did you make your biggest bubble?

    By blowing in the most air slowly.

13. How did you make your smallest bubble?

    By blowing in the least air. By blowing air in fast.

See *Developing Critical Thinking Through Science, Book Two,* for more creative bubble activities.

ACTIVITY 24: DANCING GRAPES!

Goal: To understand that air bubbles can make objects rise in a liquid

Skills: Observing, describing, inferring, predicting, explaining, sequencing, generalizing, and comparing

Materials: 2 liters of unopened club soda
1 large bunch of seedless grapes
7 transparent plastic tumblers

Preparation: 1. Divide the students into 6 groups.
2. Distribute to each group:
   1 plastic tumbler
   2–3 grapes
3. For best results, don't open the club soda until you're ready to use it—the fizz is very important!

Preparation Time: 5 minutes

Lesson Time: 10–15 minutes

— Procedure and Questioning Strategy —

Tell students to observe what you are doing. Hold up a bottle of club soda, tell the students what it is, open it, and pour some into a plastic tumbler. Hold the bottle and the tumbler of club soda so all students can see them.

1. What did you observe?
   You poured club soda into the tumbler.

2. Describe the club soda.
   It has bubbles in it. The bubbles are moving around in the club soda.

3. What do you think is in the bubbles?
   Air.

Hold up a few grapes.

4. What do you think the grapes will do when we drop them into the club soda?
   (Sink or float—predictions will vary.)

5. We're going to test our predictions in your groups. I will pour some club soda into your tumbler. As soon as I do, drop 2 or 3 grapes into the tumbler. Observe the grapes carefully for a few minutes, then discuss together what you have observed and why you think it happened. Try to talk quietly so other groups won't hear what your group says. Later we will share what we observed.

Pour club soda for each group and allow time for observation and discussion. When the groups have completed their observations and discussions, continue with the lesson.

6. What are the bubbles doing in your tumbler of club soda?

    They are moving up to the top of the water.

7. Put your hand over the top of the tumbler. Can you feel the bubbles that have gone into the air?

    Yes.

8. Why do you think they are doing that?

    There is air in the bubbles.

9. What did the grapes do when you first dropped them into the club soda?

    They sank to the bottom of the tumbler.

10. What happened after the grapes had been on the bottom for a little while?

    There were bubbles on the grapes.

11. Describe what happened next.

    The grapes moved up and down in the soda.

12. Why do you think the grapes moved up in the soda?

    There were lots of bubbles on the grapes.

13. What is in the bubbles?

    Air.

14. Do you think the air bubbles are lighter or heavier than the grapes?

    Lighter.

16. So how do you think the bubbles made the grapes go up?

    The bubbles helped them move up.

17. What happened when the grapes got to the surface (top) of the soda?

    They went back down again.

18. Why do you think that happened?

    The bubbles on top of the grapes burst when they hit the air above the soda.

19. How do you know?

    We could feel the bubbles bursting on our hands. There were fewer bubbles on the grapes.

20. Why couldn't the grapes float without so many air bubbles attached to them?

    The grapes were too heavy to float.

— Practical Application —

1. What do you use to help you float when you're swimming?

   Answers may include water wings, swim rings, and inner tubes.

2. What is similar about the things you use to float and the bubbles?

   They all have air in them.

3. How are they different from the bubbles?

   They don't burst when they're on the surface of the water, so we don't sink like the grapes do.

ACTIVITIES 19–24: CONNECTIONS

Goal: To find relationships among the activities for air

Skills: Observing, describing, using data gathered, explaining, drawing conclusions

Materials: Balloon

— Procedure and Questioning Strategy —

Blow up the balloon and hold the neck of the balloon closed.

1. Let's review what we've learned about air. How do we know that air is in a balloon that's blown up?

    We can feel it when we push on the sides of the balloon.

Have a student feel the air as you let it go out of the balloon.

2. How else do we know that there is air in the balloon?

    We can feel it when it comes out of the balloon.

Blow up the balloon again.

3. What happens when the balloon is let go?

    It flies around the room.

4. What makes it go?

    The air inside the balloon goes out.

5. Where else can we find air?

    In the room. Around us. Outside.

6. How do we know air is there?

    We caught it in bags.

7. How do we know that air takes up space?

    It took up space in the balloons and bags.

8. What about when we tried to get water into a blown up balloon?

    The water pushed the air out of the balloon.

9. How did we know that happened?

    Air bubbles came up from the balloon through the water in the funnel.

10. What about when we put the soda bottle into the water upside down?

    Just a little bit of water went up into the bottle.

11. What took up the space in the rest of the bottle?

    Air.

12. How did we know that?

    The water couldn't get in where there was air.

13. What happened when we put the cup with the paper towel in water?

    The towel stayed dry. The air in the cup took up the space. The air wouldn't let the water go up that high in the cup.

14. What about the bubbles we blew? How did we make small bubbles?

    By blowing in a lot of air very slowly.

15. How did you make big bubbles?

    We blew in more air. We turned the cup upside down.

16. Which air took up more space? The air in the small bubbles or the air in the big bubbles?

    The air in the big bubbles.

17. What made the grapes rise in the soda?

    The air bubbles were on the grapes.

18. How do we know that the air bubbles made the grapes float?

    When the bubbles burst, the grapes sank.

19. Where do we find air?

    It's all around us—in the room and outside.

20. How do we know air is there?

    It takes up space in things like balloons and bubbles. Our bag popped open when we pulled it through the air . When we closed the top of the bag, we could feel the air inside the bag.

21. What did we learn about air with the funnel and the balloon?

    Air is lighter than water. The heavy water pushed the air out. Air and water couldn't stay in the same place.

22. What did we learn about air with the cup and the paper towel?

    If the air can't get out of something, water can't get in.

UNIT 5: MOVING AIR – AIR PRESSURE

ACTIVITY 25: FANS AND PINWHEELS

Goal: To experience air that is moving
To understand that air can move things

Skills: Observing, describing, inferring, explaining, summarizing

Materials: For each student:
piece of facial tissue or 3 squares of toilet paper
piece of paper to make fan
copy of pinwheel pattern (on the last page of this lesson)
scissors
paste or small pieces of transparent tape
straight pin
pencil with eraser

Preparation: 1. Make copies of the pinwheel pattern.
2. Assemble the materials for the students and for the pinwheel making demonstration.

Preparation Time: 5 minutes

Lesson Time: 30 minutes

— Procedure and Questioning Strategy —

1. Blow on the palm of your hand. How does that make your hand feel?

   Like something's touching it.

2. Can you see anything touching your hand?

   No.

3. What do you think it is that's touching your hand?

   Air.

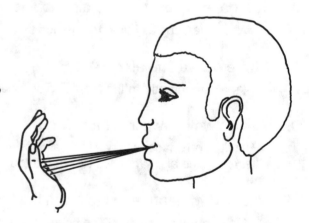

4. How do you know?

   I'm blowing the air out of my mouth and it's hitting my hand.

5. So you're moving air from your mouth to your hand.

Give each student a small piece of facial tissue or toilet paper.

6. Hold the tissue in front of your mouth and blow air toward the tissue. What happens to the tissue?

    It moves away from my hand.

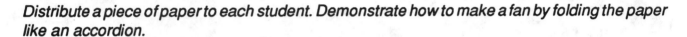

7. What happens if you blow hard?

    The tissue moves farther away.

8. What is making it move in that direction?

    The air I'm blowing is pushing it away.

Distribute a piece of paper to each student. Demonstrate how to make a fan by folding the paper like an accordion.

9. Fan your face with your fan. What do you feel?

    Air.

10. What is moving the air?

    The fan.

Demonstrate how to make a pinwheel.

11. – First we'll cut out the whole pinwheel square.
    – Then we'll cut each diagonal line until we reach the outside edge of the black dot in the middle of the pinwheel. We need to be careful not to cut through the dot because the dot has to hold the arms of the pinwheel together.
    – Notice that the corners of the pinwheel pattern are numbered 1–4.
    – We'll pull corner 1 to the center of the pinwheel and paste it (or tape it with a tiny piece of tape), being careful not to crease the bend.
    – We'll repeat this same procedure with corners 2, 3, and 4.
    – Now we'll push the straight pin through all four corners in the middle of the pinwheel until it comes out the other side.
    – Then we'll push the pin firmly into the eraser on the end of the pencil. We'll make sure to leave enough room between the pencil and the pinwheel blades so that the pinwheel can move freely.
    – We'll give the paste a few minutes to dry before using the pinwheel.

Distribute the materials to the students. Caution them to be careful with the pins.

12. Now make your own pinwheels.

Give the students a few minutes to complete their work. Assist students who are having difficulty. If many students are having problems, demonstrate making another pinwheel while students follow along making their own.

13. Blow on your pinwheel. What happens?

   It moves around. It turns.

14. What is making it move?

   The air from my mouth.

15. Yes, when you blow air, or wind, it makes the pinwheel move.
   Can you find another way to make it move?

   (Students may use their fans to make the pinwheel move. They may
   wave their pinwheels back and forth or walk with them holding them forward.)

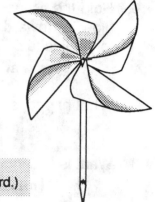

— Practical Application —

Take the students outdoors to try their pinwheels. Have them try to move their pinwheels by blowing on them, holding them facing the breeze or wind, and running with them.

Pinwheel Pattern

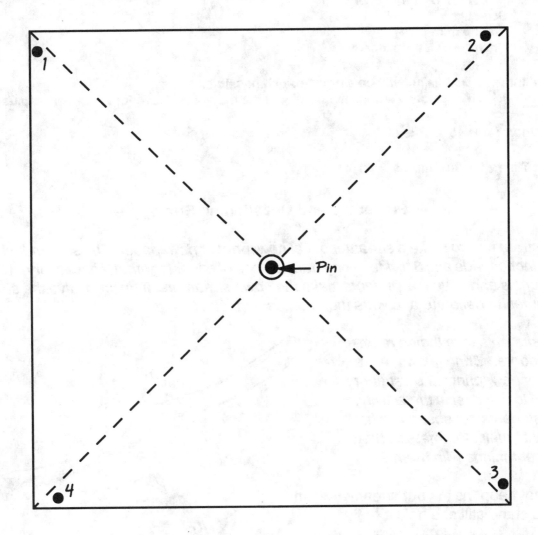

ACTIVITY 26: STREAMERS

Goal: To understand that moving air helps things move and holds things up

Skills: Observing, describing, comparing, inferring, explaining

Materials: Newsprint or newspapers
Scissors
Transparent tape
Crayons or felt-tip markers

Preparation: 1. Do this activity on a breezy day if possible.
2. Assemble a sufficient quantity of the materials above for all students to use.

Preparation Time: 5 minutes

Lesson Time: 25–30 minutes

— Procedure and Questioning Strategy —

Demonstrate how to make a streamer out of newsprint or newspaper. The streamer should be about 2 inches wide and 3 to 4 feet long (tape pieces of paper together if necessary). Distribute the materials and have the students make their own streamers. If desired, give the students a few minutes to decorate and write their names on their streamers.

Let the students take their stream-ers outdoors. Bring along extra lengths of newsprint or newspaper and tape to use later in the activity. Guide students to observe what happens to their streamers as they do different things with them.

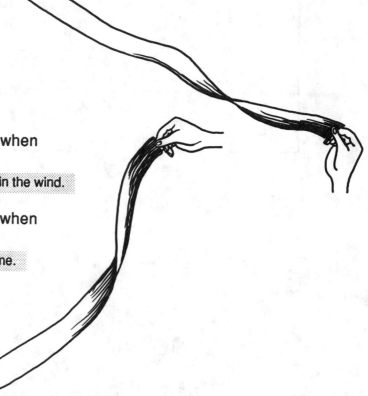

1. What happens to your streamer when you stand still and hold it up?

    It stays still. It moves a little (a lot) in the wind.

2. What happens to your streamer when you run with it?

    It flies. It goes straight out behind me.

3. What does the streamer look like when the wind blows it?

    Like a flag. It goes straight out.

If there is a breeze, attach a streamer to an object like a piece of playground equipment. Have students observe.

4. In which direction does the streamer move?

    It hangs down. It waves toward the school, etc.

5. What do you think is making it move?

    Air. The wind.

6. How does it move when there is a breeze?

    It flutters. It makes noise. It waves. It moves fast.

7. How does it move when the wind blows harder?

    It goes straight out. It wants to fly away. It has little waves.

Have a student fasten a streamer to a swing and have another student swing back and forth so the streamer moves with the swing. Have the students add more length to their streamers and compare movements with shorter streamers. When students are back in the classroom, discuss what they have observed and their explanations.

8. What happened to your streamer when you stood still and held it up?

    It hung down.

9. Did it always hang down?

    No. Not when there was a breeze.

10. What did it do in the breeze?

    It waved a little.

11. What happened to your streamer when you ran with it?

    It blew out behind me and flapped a lot.

12. What did the streamer look like when the wind blew?

    It blew out and flapped in the wind.

13. What about when just a little breeze blew?

    It moved up a little and fluttered.

14. What happened when a streamer was attached to the swing and (student's name) made the swing move back and forth?

    The streamer lifted up and waved. It moved back and forth with the swing. It moved one way when the swing went up and the other way when the swing went back.

15. What do you think lifted the streamers up and made them wave and flutter when you ran with them?

    The air.

16. How do you know this? Can you think of something we did on another day that reminded you of the streamer blowing in the breeze?

    Yes, when we blew the tissue that we were holding in our hand it lifted up like the streamer does in the wind.

17. What lifted them up and made them move when you were just holding them up?

    The wind or breeze (moving air).

— Practical Application —

1. What other things have you seen that were moved by the wind?

    (Possible answers: kites, flags, wind socks, weather vanes, leaves, papers, dust, hats, umbrellas, clothes, trees, grass, sand.)

ACTIVITY 27: PARACHUTES

Goal: To understand that air holds up objects
To understand that air makes parachutes fall slowly because it helps hold them up

Skills: Observing, comparing, classifying, describing, explaining

Materials: 2 handkerchiefs or squares of lightweight cloth
4 pieces of string, each about 14" long
2 large paper clips
Several metal washers
Thread

Preparation: 1. Make two parachutes. (See the illustration on the next page.)
2. Use the thread to tie together 3 of the washers.

Preparation Time: 15 minutes

Lesson Time: 30–35 minutes

HOMEMADE PARACHUTES

1. Tie the ends of two pieces of string to the opposite corners of a handkerchief.

2. Hang a large paper clip where the two strings cross as they hang down.

3. Hang a washer from the bottom of the paper clip.

— Procedure and Questioning Strategy —

Place the washers and parachutes where you can reach them easily. You may wish to choose a student volunteer to retrieve washers after they have dropped to the floor. Stand on a stool or chair. Hold up a separate washer so the students can see it.

1. What do you think will happen if I drop this washer?
   It will fall down to the floor.

Drop the washer.

2. What happened?
   The washer fell down. It hit the floor.

Hold up a single washer and the three washers tied in a group.

3. Which do you think is heavier—the single washer or the three washers tied together?
   The three washers tied together.

4. What will happen if I drop the single washer and the three washers tied together at the same time?
   (Students usually say that the three washers will fall faster or hit the floor first.)

Drop the single washer and the three tied washers, making sure that the bottoms of the washers are even before you drop them.

5. What happened?

   They hit the floor at the same time.

6. So even though the washers are different weights, they hit the floor at the same time?

   Yes.

Hold up one of the parachutes with a washer attached.

7. What do you think will happen if I drop the washer attached to the parachute?

   It will drop down to the floor. The parachute will open up.

8. Let's watch it.

Hold the parachute from the middle of the handkerchief, then drop it.

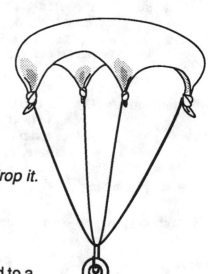

9. What happened?

   The washer dropped to the floor. The parachute opened.

10. Did this washer fall as fast as the ones that weren't attached to a parachute?

   No. It looked like it was slower.

Hold up a single washer and a parachute so that the bottoms of the washers are at the same level.

11. Let's drop a washer by itself and a washer attached to a parachute at the same time so we can compare them.

At the same time, drop the separate washer and the parachute with the washer attached. Make sure the bottoms of the washers are at the same height when you let them go.

12. What happened when we dropped both washers at the same time?

   The one on the parachute fell slower.

13. Why do you think the washer on the parachute fell slower?

   The parachute opened up.

14. Why did the parachute open up?

   There was air under it.

15. What do you think the air did to the parachute—what made it go slower?

     The air helped hold it up.

Hold up the two parachutes with the single washers attached to them. Make sure you hold them so that the bottoms of the washers are at the same height.

16. What do you think will happen if I drop both parachutes at the same time from the same height?

     They will hit the floor at the same time.

17. Let's try it.

18. What happened?

     They did hit together.

If this does not happen, adjust the parachutes so they are as similar as possible and try it again.

19. I'm going to add two more washers to one parachute. What do you think will happen if we drop the two parachutes again?

     The one with three washers will drop faster.

Drop the parachutes.

20. What did you observe?

     The one with three washers hit the floor first.

21. Did adding more washers make a difference?

     Yes.

22. Why do you think it made a difference?

     The parachute with three washers was heavier.

23. What happened when we dropped one washer and three washers at the same time when they weren't attached to parachutes?

     They hit the floor at the same time.

24. But when we dropped three washers attached to a parachute and one washer attached to a parachute, the three washers fell faster. Why was that?

     There was enough air under the parachute to slow down one washer but not enough to slow down three.

25. What can we conclude from our experiment?

     Air holds the parachute up, but if the washers are heavier, the air can't hold them up as well.

26. Let's test our conclusions a few more times to be sure.

Drop the parachutes a few more times to give students more evidence for their conclusions.

— Practical Application —

1. How are these parachutes like the ones people use when they go skydiving?

   They're both made of cloth that opens up. They both have strings/cords attached to the cloth. They both have something or someone heavy hanging at the bottom of the strings or cords.

2. How is the type of parachute that skydivers use different from the type we're using?

   The skydiving parachute is a lot bigger and stronger. It has cords instead of strings. It is attached to a harness on a skydiver's body instead of to a washer. A skydiver pulls a cord to open the parachute as s/he is falling. Our parachute opens by itself when it's dropped.

3. Imagine that two people with the same size and kind of parachute jump out of a plane. One person is much bigger and heavier than the other person. Both people pull their parachute cords at the same time. Both parachutes open at the same time. Will both people land on the ground at the same time, or will one person get there first?

   The heavier person will land first.

4. Why do you think that will happen?

   The parachute can't hold a heavier person up as well as a lighter person.

5. How do you know that?

   Our parachute with the three washers hit the ground first.

* * * * *

Make parachutes of different sizes. Attach the same number and different numbers of washers to each of them. Test to see if there is a difference in the speed of the parachutes.

ACTIVITY 28: PAPER AIRPLANES

Goal: To understand that air helps hold up paper airplanes so they can fly
To explore ways to design paper airplanes that change the way they fly

Skills: Observing, planning, designing, testing, redesigning, discussing, comparing, explaining, generalizing, evaluating results of experiments

Materials: 8 1/2" x 11" sheets of paper, several per student

Preparation: Practice making the paper airplane for the demonstration.

Preparation Time: 5–10 minutes

Lesson Time: 30–40 minutes

— Procedure and Questioning Strategy —

Demonstrate for the students one possible way to make a basic paper airplane using the directions on the last page of this activity. As you make the airplane, describe what you are doing, i.e., folding, creasing, lining up edges, folding up or under, etc. Go over the names of the parts: wing, nose, tail, etc.

Fly the airplane a few times then ask students the following questions.

1. How would you describe the flight of the airplane?

   (Answers vary.)

2. What made the airplane fly?

   The way you held it. The way you threw it. The way it was made.

3. What helped hold it up?

   The air under it.

4. How did the plane land?

   (Answers vary.)

5. What do you think we could do to make the plane fly farther?

   Throw it harder. Hold it higher when throwing it. Make it another way.

Distribute several sheets of paper to each student. Have the students make their own airplanes as you demonstrate the step-by-step procedure again and give guided practice.

Chose a suitable area for testing the airplanes. Since classrooms are rather limited in space and have many obstacles, a gymnasium, multi-purpose room, or playground can be used. If *the day is windy, an indoor testing area is preferable. When the students have finished testing their airplanes, discuss the reasons they think some planes flew farther than others. Possible reasons: the paper airplane had straighter or harder creases; it had a nose that was more pointed; the plane was thrown harder, straighter, or from a higher position.*

— Practical Application —

Give the students a chance to design their own airplanes that can fly even farther. They can start with one like they have made or change the design in some way. Let the students work alone or in pairs or teams. Teamwork is important for those students who would benefit from help in design and construction. Encourage partners to explain to one another why they are making changes in design.

Plan enough time to test the airplanes. Have extra paper available so students can change the design of their planes after the first test. Ask the students to explain what was changed and how these changes affected the flight distance. Allow time for redesigning and retesting. Guide the students to make generalizations. For example: the harder the airplane is thrown, the farther it goes; when the airplane is thrown with its nose pointing up, it flies upward, then curves down; airplanes with wider wings have shorter flights.

* * * * *

*Schedule a special day for an **"Air Show and Celebration."***

Invite students to design an airplane that
> *– can make a soft landing*
> *– can be controlled to land in a particular area, like an*
>   *outlined circle, from a specific launching area*
> *– has the longest, straightest flight*
> *– can curve in the air while flying*
> *– has the most unusual design (and can still fly)*
> *– has the most beautiful decoration*

Add the students' and your own special categories for airplane design.

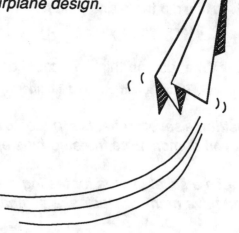
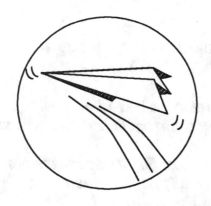

Basic Paper Airplane

1. Fold the paper in half lengthwise.

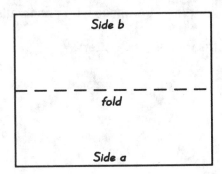

2 Fold each side down three times.

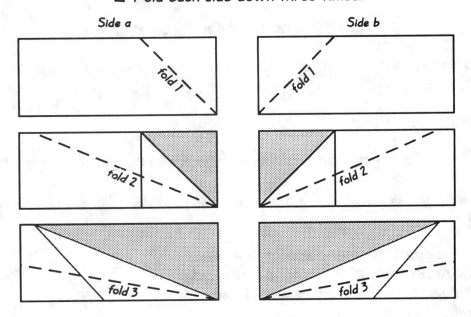

3. The top view of the finished plane should look like this.

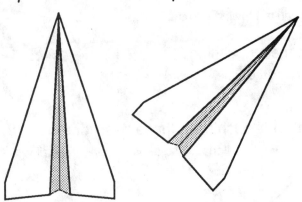

ACTIVITY 29: THE STRENGTH OF AIR

Goal: To understand that air is strong
To understand that the pressure exerted by air can lift objects and hold them up

Skills: Observing, describing, supporting inferences with observations, summarizing

Materials: Large, round balloon (the thicker, the better)
Books
Rubber bands
Empty jar, about 16 oz.

Preparation: None

Preparation Time: None

Lesson Time: 15–20 minutes

— Procedure and Questioning Strategy —

1. We have found that air can move things and hold them up. Let's try some other experiments with air.

Show the students the balloon and a book.

2. Can you think of a way that we can lift up this book with the balloon without touching the book with our hands?
    (Answers vary. One answers may be to put the balloon under the book then blow up the balloon.)

Try the students' suggestions, then place the balloon flat on a table so the neck of the balloon hangs over the side of the table. Put a book on top of the balloon. Blow into the balloon to lift the book.

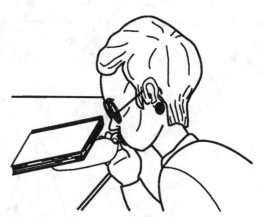

3. What is lifting the book?
    The balloon. The air you're blowing into the balloon.

4. In which directions is the air pushing?
    Toward the book. Toward the table. Toward the inside of the balloon.

5. How do we know that air is pushing in those directions?
    If it weren't, the balloon would collapse. The air is pushing the balloon out against the table and the book.

Try the experiment again using two or three books held together by rubber bands. Discuss what happened.

6. Did adding more books make lifting more difficult?

    Yes.

7. Why do you think that adding more books made lifting harder?

    The two (three) books were heavier than one book.

8. So, what have we found out about air?

    It is strong. It can help lift things. It takes more air to lift heavier things.

— Practical Application —

1. Think about a car. What holds the car up?

    The tires.

2. What if the car had a flat tire? What would have to be done to the tire?

    It would need to be fixed. It would have to have air put back in it.

3. As the air is put back into the tire, what happens to the rest of the car?

    It's lifted up.

4. How is that like what we did with the balloon and the books?

    The air that is being blown in helps to lift the books and the car.

5. So we have found that air is strong. To lift something as heavy as a car, air has to be blown into something like a tire that is strong enough not to burst.

* * * * *

Show the students the jar and the balloon.

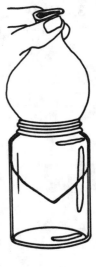

1. Can you think of a way to lift this jar with the balloon—without touching the jar?

    (Answers vary. One may be to put the balloon into the jar and blow it up until it can lift the jar.)

Try each suggestion then hold the balloon so it hangs partly into the jar. Blow it up until you can lift the jar by just holding and lifting the neck of the balloon.

2. What is holding up the jar?

    The balloon.

3. Could the balloon hold up the jar all by itself?

   No. It needs to have air inside it.

4. What does this tell us about the air inside the balloon?

   It's strong. It's pushing against the sides of the balloon.

5. Could the air hold up the jar without using the balloon?

   No. The air would spread out all over.

6. So it takes both the balloon and the air inside of it to hold up the jar.

7. In which directions is the air inside the balloon pressing?

   Against the sides of the jar.

8. How do you know that?

   The balloon is indented (pushed in) where it is pressing on the jar.

9. Is the air pressing in any other direction?

   Against the sides of the balloon that are not in the jar.

10. Why do you think the part of the balloon that's in the jar is not blown up as much as the rest of the balloon?

    (Answers vary.)

11. What was in the jar before we blew up the balloon inside it?

    Air.

12. Where did the air go when the balloon pressed hard against the sides of the jar and blew up into the jar?

    It was pushed to the bottom of the jar.

13. So what is keeping the balloon from blowing up more inside the jar?

    The air trapped at the bottom of the jar. It is pushing hard against the bottom of the balloon.

ACTIVITIES 25–29: CONNECTIONS

Goal: To show the relationships among the activities for moving air and air pressure

Skills: Comparing, classifying, summarizing

— Questioning Strategy —

1. We have found that air is strong. It can move things and help hold things up.

2. What did we move with air?

   Pinwheels, streamers, the jar, and books.

3. What did we do that showed air could help hold things up?

   Air helped hold the parachutes up. We flew airplanes. We lifted up the jar and books with the blown up balloon.

4. Did we have to move air to hold up the jar and the books?

   Yes. We had to move air from our mouth into the balloons.

Take the students on a walk to see if they can observe signs of moving air (for example: flag, leaves, tree branches, paper blowing in the wind). Do they hear the wind moving? To help students remember what they've seen and heard, they might carry a notepad and pencil or crayon to write or draw quick sketches.

* * * * *

Have them look up and describe the movement of the clouds. Tell them that wind is moving the clouds and help them discover which direction the clouds are moving.

* * * * *

When everyone is back in the classroom, ask the students to name the things that were being moved by wind. List them on the board or on an overhead. Have the students share the pictures they drew or what they wrote about when they were outside.

1. All of these things we listed, as well as our pinwheels, needed moving air or wind pushing them to make them move. Can you think of any other things that need air to move or work?

   (Students might mention kites, sailboats, balloons, paper airplanes, real airplanes, hot air balloons, windmills, wind instruments—like trumpets, horns, saxophones—vacuum cleaners.)

UNIT 6: FORCE

═══ ACTIVITY 30: PUSH AND PULL ═══

Goal: To understand that force is a push or a pull

Skills: Observing, estimating, comparing, using observations to support inferences

Materials: 3 or 4 classroom objects different enough in weight so that students can easily order (classify) them from lightest to heaviest, for instance:
rubber band
pencil
box of paper clips
box of crayons
books of different weights

Preparation: Put the objects on a table where all students can see them.

Preparation Time: 5 minutes

Lesson Time: 25–30 minutes

— Procedure and Questioning Strategy —

Have a couple of students come up to the table and lift the objects one at a time.

1. What differences do you feel when you lift the objects?
   Some are heavier (or lighter) than others.

2. How do you know that?
   Some are harder (or easier) to lift than the others.

3. Decide together how to arrange these objects in order from lightest to heaviest.

When the objects are arranged in this order, ask two other students to lift each object, in order from lightest to heaviest, to find out if they agree with the classification.

4. When you move something by lifting it, you are using force. A force is the strength used to lift something up, move it toward you, or push it away from you.

Demonstrate these movements.

5. When did you use the most force (strength) to lift an object?
   When I lifted the (name of heaviest object).

6. How do you know that?

   It was the hardest to lift.

Have two other students push each object across the table.

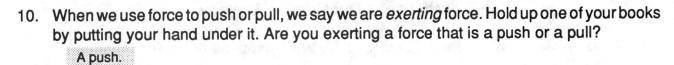

7. Which was the hardest to push?

   The (name of heaviest object).

8. Which was the easiest to push?

   The (name of lightest object).

9. Which object did it take the most force to push?

   The (name of heaviest object).

10. When we use force to push or pull, we say we are *exerting* force. Hold up one of your books by putting your hand under it. Are you exerting a force that is a push or a pull?

    A push.

11. In which direction are you exerting your force?

    I'm pushing upward, against the book.

12. Is the book exerting a force against your hand?

    Yes.

13. How do you know that?

    It feels heavy. It's pushing down against my hand.

14. What do we call a push or a pull?

    A force.

15. When we use force, what do we say we're doing?

    Exerting a force.

— Practical Application —

Have the students name objects at home, school, or elsewhere, or things they use, that they either push or pull. Make two lists, one for PUSH and one for PULL. For example:

PUSH — Doorbell button, computer keys, bike pedals, instrument keys, door to close
PULL — Rope in tug of war, venetian blind or curtain cord, lawn mower starter cord, light or ceiling fan cord, drawer to open

*Some objects can be included in both lists, like drawer listed above. Make sure students name some things that can **only** be pushed and some that can **only** be pulled.*

ACTIVITY 31: EARTH-PULL (GRAVITY)

Goal: To understand that the earth exerts a pulling force on objects called earth-pull (gravity)

Skills: Observing, hypothesizing, describing, inferring, supporting inferences with observations

Materials: Chalkboard eraser
Books

Preparation: None

Preparation Time: None

Lesson Time: 15–20 minutes

— Procedure and Questioning Strategy —

1. Watch while I hold the eraser, then let it go. What happened?

   It fell to the floor.

2. Why do you think it fell to the floor instead of going up or sideways?

   (Answers vary.)

3. The earth we live on is exerting a pulling force on the things on the earth. It is called earth-pull or gravity.

Drop the eraser again.

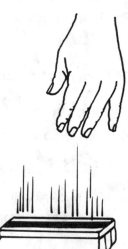

4. Why did the eraser fall to the floor?

   The earth-pull (gravity) made it fall.

5. Is earth-pull a force?

   Yes.

6. Can we see it?

   No.

7. How do we know the force is there?

   The eraser was pulled down to the floor.

8. If you jump up in the air, what happens?

   We come back down again.

9. Why don't you stay up in the air?

   Earth-pull (gravity) pulls us down.

— Practical Application —

1. Hold up one of your books with your hand under the book. Is the book exerting a force on your hand?

    Yes.

2. How do you know that?

    It's pushing down on my hand.

3. How do you know it's pushing down on your hand?

    It feels heavy.

4. What kind of force are you using to hold up the book—a push or a pull?

    A push.

5. If you took your hand away, what would happen to the book?

    It would fall down. It would fall on the floor.

6. What would make it fall down?

    Earth-pull (gravity).

7. So let's think about the forces around the book. What is the force that's pulling down on the book?

    Earth-pull (gravity).

8. What is the force that is pushing up on the book?

    My hand.

9. Let's draw a picture of the forces that are being exerted on the book. In which direction should we draw an arrow to show earth-pull force?

    Down.

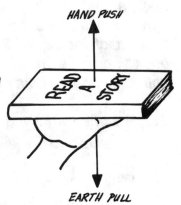

HAND PUSH

EARTH PULL

10. In which direction should we draw an arrow for the pushing force of your hand?

    Up.

ACTIVITY 32: HOW MUCH FORCE DOES IT TAKE?

Goal: To develop an understanding of force

Skills: Observing, measuring, describing, comparing, explaining, graphing, supporting inferences with evidence

Materials: A school book like one that all students have in their desks
String
Rubber bands
Large paper clips
Spring scales marked in Newtons to measure force
   0–20 Newton scale, 0–50 Newton scale (Most schools have these, or they can be purchased at school supply stores.)
Measuring tape if a spring scale is not available
Graph on the last page of this lesson—one copy for each student

Preparation: None

Preparation Time: None

Lesson Time: 20–25 minutes

— Procedure and Questioning Strategy —

Have the students watch as you slip a piece of string through the middle pages of a book and tie it loosely outside the binding edge of the book. Make a chain of four paper clips and three rubber bands alternating the paper clips and rubber bands. Hold the chain up for the students to see. Now attach the paper clip at one end of the chain to the string outside the book binding. Hold the clip at the other end of the chain. Lift the book slowly until it is hanging from the chain.

1. Describe what you see here.

   (Accept all answers that describe what the students observe.)

2. Why is the book hanging down?

   The earth is exerting a downward force on it called earth-pull.

3. Look closely at the rubber band and paper clip chain. Notice how long it is. I am using force to hold the book up and gravity is the force that is trying to pull it down. Both forces are shown by the length of the rubber band and paper clip chain.

4. Now, I'm going to put the book on the end of this table. Then I'm going to pull it along the table. Notice how long the rubber band and paper clip chain becomes as I pull the book.

Pull the book slowly across the table with the chain as close to the table as possible.

5. When was the rubber band and paper clip chain shorter? When I held the book up or when I pulled it across the table?

    When you pulled it across the table.

6. What does this tell us about the force used to pull the book?

    It took less force to pull it.

7. Why do you think it took less force?

    The table helped hold the book up.

8. So it took more force to hold the book up than to pull it across the table.

9. We can measure this force on a special scale. When we want to measure our weight, we weigh ourselves on a scale that measures in units called pounds. When we want to measure force, we use a spring scale like this one. A spring scale measures in units called Newtons. Each of the marks on this spring scale indicates one Newton.

Demonstrate for the students. Take the rubber band and paper clip chain off the book and attach a Newton spring scale to the string. Hold the book up with the spring scale. Read the Newtons of force on the spring scale. (If a spring scale is not available, have students measure the rubber band and paper clip chain with a measuring tape.)

10. It takes _____ Newtons of force to hold up the book. Do you think it will take more or less Newtons of force to pull the book across the table?

    Less Newtons.

Pull the book along the table with the spring scale. Have one of the students read the scale to see how much force is being exerted to pull the book.

11. How much force did it take to pull the book?

    _____ Newtons.

12. So the spring scale showed that it did take less force to pull the book across the table than to hold it up.

13. In your desks, you have a book like this one that we've been using to help us measure force. Take it out and hold it up. It takes you as many Newtons of force to hold up your book as it took the spring scale to hold it up. How many Newtons was that?

    _____ Newtons.

14. So you are using _____ Newtons of force when you hold up your book.

— Practical Application —

Use the spring scale to pull various things in the room like a venetian blind or curtain cord, a projector screen or map cord, and a drawer. In some way, mark each object with the amount of force needed to pull it. Graph the results on a copy of the bar graph provided at the end of this lesson. A completed sample graph is shown at right.

* * * * *

Allow the students to use the spring scale to explore the amount of force needed to lift other objects in the room so they get a feel for the amount of force needed. Graph some of the forces used to hold or pull some classroom objects so the forces can be compared.

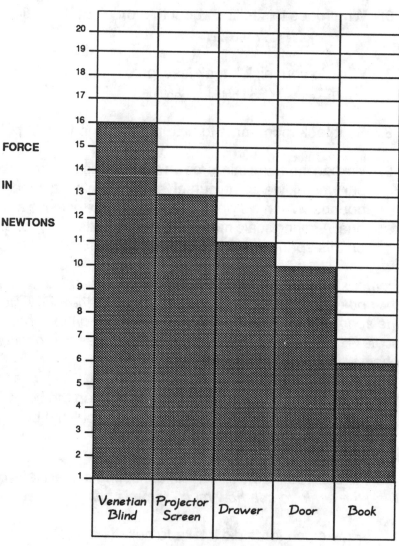

SAMPLE BAR GRAPH

HOW MUCH FORCE DOES IT TAKE?

FORCE IN NEWTONS

OBJECTS MEASURED

HOW MUCH FORCE DOES IT TAKE?

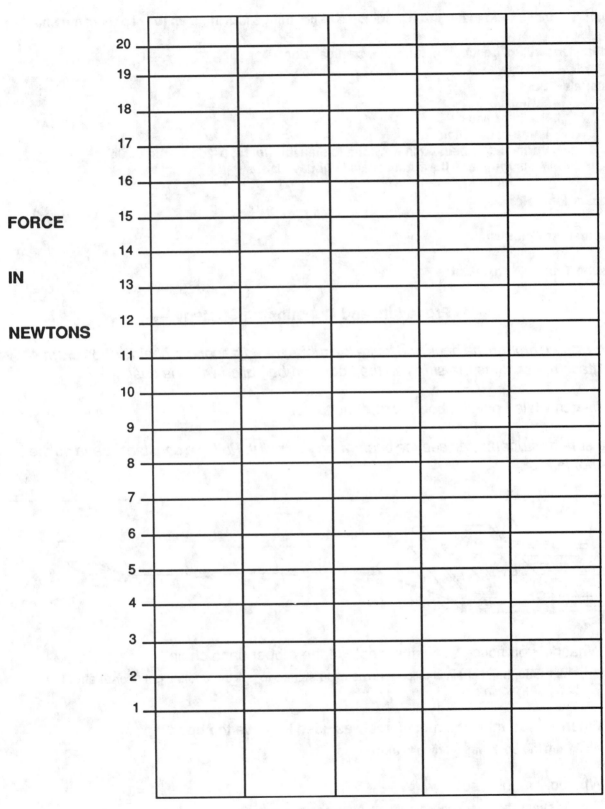

FORCE

IN

NEWTONS

OBJECTS MEASURED

— ACTIVITY 33: GETTING THINGS MOVING, KEEPING THEM MOVING —

Goal: To understand that it takes more force to get things moving than to keep them moving

Skills: Observing, describing, inferring, summarizing

Materials: Book
String
Rubber bands
Large paper clips
Spring scale marked in Newtons to measure force, 0–20 Newton scale
Measuring tape if a spring scale is not available

Preparation: None

Preparation Time: None

Lesson Time: 15–20 minutes

— Procedure and Questioning Strategy —

Tie the string through the book and make the rubber band/paper clip chain as you did in Activity 32. Attach the chain to the string so the book can be pulled with the chain.

1. Watch while I pull the book across the table.

Make sure all students can see the book when you pull it. Pull at a constant speed so the book does not jerk.

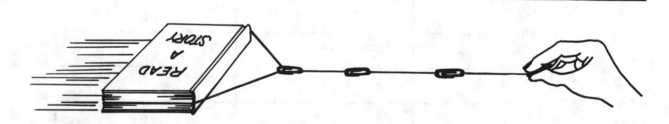

2. What did you notice about the length of the rubber band chain?
   The rubber band chain was longer when you started to pull the book. It got shorter after the book was moving.

3. When do you think the most force was used to move the book?
   When the book was starting to move.

4. Why do you think so?
   The chain of rubber bands was the longest then.

5. Let's measure the force in Newtons using the spring scale (or use the measuring tape to check the difference in chain length).

Remove the rubber band chain and replace it with the Newton spring scale. Have a student read the scale as you pull the book across the table.

6. How many Newtons of force did it take to start the book moving?

   _____ Newtons.

7. How many Newtons of force did it take to keep the book moving?

   _____ Newtons.

8. Which took more force?

   Getting the book started.

9. Who can tell us what we've found out about force?

   It takes more force to start something moving than to keep it moving.

— Practical Application —

1. Think about the force you use when you ride your bicycle. When do you have to push down on the pedals the hardest?

   (Answers vary but they should include getting the bike started.)

2. When do you use the most force when you are swinging on a swing?

   When you first get the swing started—you have to pull really hard.

3. What about on a merry-go-round—when do you have to use the most force?

   When you first get the merry-go-round started.

4. So those are the times when you have to use a lot of force. When do you use the least force to do these things?

   (Answers should include: when the bike is already moving; when the swing is already swinging or swinging highest; when the merry-go-round is going around fast.)

5. How do these activities compare with what we found out about force when we pulled the book across the table?

   It takes more force to get something started than to keep it moving.

6. So, what have we learned about force or the strength needed to move an object?

   It takes more force to start something moving than it does to keep it moving.

ACTIVITY 34: FORCE OF FALLING OBJECTS

Goal: To understand that the farther an object falls, the more force it has when it lands

Skills: Observing, comparing, classifying, inferring, using evidence from observations to support inferences

Materials: For demonstration:
   shallow container like a cake pan
   beach or builder's sand (or very soft modeling clay)
   2 marbles, the same size and weight

Preparation: None

Preparation Time: None

Lesson Time: 15–20 minutes

— Procedure and Questioning Strategy —

Have the students observe as you spread and smooth out the sand or clay in the bottom of the container. Put the container on the floor where students can see it. Hand two marbles to a student.

1. Look at these two marbles and tell us how they are alike.

   They are the same size and shape.

2. Is one marble heavier than the other?

   No, they seem to be about the same weight.

3. Tell us how they are different.

   (Answers will vary.)

4. Watch while I drop one of the marbles into the container. I'm going to drop it from about waist high.

After you drop the marble into the sand, carefully pick up the container and let the students see how deep the marble sank.

5. This shows the amount of force the marble had when it hit. Look how deep it went into the sand.

6. What do you think will happen to the sand when the other marble is dropped from above my head?

   (Answers vary.)

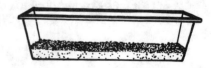

Drop the second marble into the container from as high as you can above your head. Show the students how deeply it sank into the sand.

7. What happened?

> The marble that was dropped from above your head sank deeper than the other marble. It made a wider hole (or indentation) in the sand. More sand flew up when the marble was dropped from above your head.

8. Which marble exerted the most force on the sand?

> The second marble.

9. Which marble dropped the greatest distance?

> The second marble.

Smooth out the sand and test the same marbles in the same way a few more times. Have students make observations each time the marbles are dropped.

10. Who can tell us what we have found out about the force of falling objects?

> The greater distance they fall, the more force they have when they hit.

11. Why did the marbles fall down when we let them go?

> The earth-pull (gravity) was pulling them down.

— Practical Application —

When you go outside on the playground, have the students practice jumping up in the air, making low and high jumps. Discuss the differences in force that they feel between these jumps.

1. Did you land with more force when you made low jumps or high jumps?

> High jumps.

2. How do you know that you landed with more force?

> I landed harder. It hurt my feet.

When you return to the classroom, discuss what the students have learned. Talk about safety when jumping from higher places.

3. Imagine that you are at the top of the jungle gym and you wanted to get down on the ground. What do you think it would feel like if you jumped from that height?

It would hurt your feet and legs.

4. Yes, you might seriously injure yourself. Why do you think this might happen?

    You'd land with lots of force from a place that high.

5. What would be a safer way to get down from the top of the jungle gym?

    To climb down. To climb down to a much lower bar and jump from there.

6. Why would that be safer?

    You wouldn't land with as much force.

7. Why not?

    You wouldn't be jumping as far.

8. So what did we learn this time about force and distance?

    That the farther something falls, the more force it exerts when it lands.

ACTIVITY 35: FORCE OF ROLLING OBJECTS

Goal: To understand that the farther an object rolls down an incline (slope), the more force it exerts when it lands

Skills: Observing, comparing, classifying, inferring, using evidence from observations to support inferences

Materials: For each pair of students:
ruler with groove
marble
two or three books
piece of chalk
3" x 5" index card

Preparation: Gather the materials and put them where students can reach them.

Preparation Time: 5 minutes

Lesson Time: 15–20 minutes

— Procedure and Questioning Strategy —

Divide the students into pairs. Explain that they are going to build an incline ramp to test the amount of force a rolling marble exerts when it lands. Demonstrate a way for the students to set up their ramps.

1. – Find a flat place on a table (or use the floor if it's easier).
   – Put the books on the table in a stack.
   – Place the ruler so that the end that reads "12"" rests on the top edge of the book stack and the other end rests on the table.
   – Fold the card so it looks like a tunnel.
   – Place this "tunnel" on the table with one of its folded sides against the ruler where the ruler touches the table. The tunnel should be centered on the ruler so the two objects make a "T" shape.

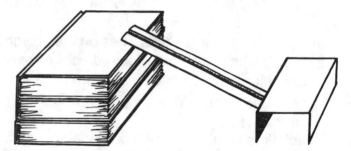

2. For each test, we're going to let the marble roll down the groove in the ruler. First, release the marble from the 1" mark near the bottom of the ruler. Use the chalk to mark on the table how far the marble pushed the index card.

3. Put the card back next to the bottom of the ruler. Let the marble go from the 2" mark on the ruler. Mark how far the card was pushed.

4. Continue your tests at each inch marking from 3" to 11". Be sure to mark how far the card was pushed each time.

Have the students gather their materials and build their ramps. Walk around among the students helping anyone who is having difficulty getting started. Remind the students to test the force of their rolling marbles by starting them at different inch markings on their ramps. When the students have completed their testing, discuss their findings.

5. What did you find out?

    The card didn't move as far when the marble rolled from the bottom of the ruler. The card moved the farthest when the marble rolled from the top of the ruler.

6. What caused the marble to roll down the ramp?

    Earth-pull (gravity).

7. When did the marble exert the most force on the card?

    When it rolled from the top of the ruler.

8. How do you know that?

    It moved the card the farthest.

9. So when the marble rolled from the top of the ramp, it moved the card the farthest. What does this tell us about force?

    The farther the marble rolls down the ramp, the more force it exerts on the card.

— Practical Application —

1. Imagine that there are two slides on the playground. One slide is twice as long and twice as high as the other slide. You slide down the shorter, lower slide then you slide down the longer, higher one. Which time would you exert more force when you land?

    When we slide down the longer, higher slide.

2. Which time would you feel your landing was harder?

    From the longer, higher slide.

3. What was pulling you down the slide?

    Earth-pull (gravity).

ACTIVITIES 30–35: CONNECTIONS

Goal: To find relationships among the activities for force

Skills: Observing, comparing, describing, inferring, applying

Materials: Box of books that isn't too heavy for a student to pick up and carry
Flimsy cardboard box

— Procedure and Questioning Strategy —

1. Let's see if we can connect some of the things we've learned about force.

2. If you were asked to move a box of books across the room, how would you do it to use the least amount of force?

    Pull it (or push it).

3. What would take the most amount of force?

    To pick it up and carry it across the room.

4. How do you know that picking it up would take more force than pulling or pushing it?

    Because it took more force to pick up a book than it did to pull it.

Put the box of books at one end of a table. Have a couple of students, one at a time, pull the box across the table then pick it up and carry it across the room.

5. Compare the forces you just used. Which took the most force—pulling the box or picking it up and carrying it?

    Picking it up and carrying it took the most force.

6. Imagine that you are sitting on a swing in a playground. You and the swing are not moving. What is the force that is holding you up?

    The swing seat, the chains attached to the seat, and the bar at the top of the swing set.

7. What is the force that is pulling you down?

    Earth-pull (gravity).

8. Is a force only a pull?

    No. It can also be a push.

9. Are you pushing on anything when you are sitting still on the swing?

    Yes. I push down on the seat when I sit on it.

This concept can be demonstrated by having a student squash a flimsy cardboard box by sitting on it. Relate the student's push on the box to a person's push on the swing seat.

10. Imagine you are sitting on a swing. Would you need to exert more force to get started swinging or to keep swinging once you were in motion?

    It would take more force to get started.

11. Is force a push or a pull?

    It's both.

12. Do you push or pull when you swing?

    Both.

* * * * *

Let the students try out their ideas about force when they go out on the playground. Discuss safety measures before leaving the classroom.

UNIT 7: SPACE, LIGHT, AND SHADOWS

ACTIVITY 36: POSITION IN SPACE

Goal: To understand and apply relative positions in space

Skills: Observing, listening, relating, comparing, applying, describing

Materials: For each student:
        small box with lid (like a school box or shoe box)
        set of small classroom objects (like a set of blocks or
           plastic manipulatives)

Preparation: Gather the materials and have them easily accessible.

Preparation Time: 5 minutes

Lesson Time: 25–30 minutes

— Procedure and Questioning Strategy —

Lead the students in a game of "Follow Me" so they can use their bodies to demonstrate positions in space. Lead them through the positions slowly at first, then faster as they get more familiar with position words. Emphasize the position words (in bold) as you give the directions.

1. Stand up with your hands **by** your sides.

2. Lift your hands **above** your head.

3. Then put your hands **on** your head.

4. Take your hands **off** your head.

5. Then lift your hands **over** your head.

6. Put one hand **between** your knees.

7. Put your other hand **in back of** your knees.

8. Hold both hands **in front of** your waist.

9. Hold one hand **on top of** the other hand.

10. Put one hand **behind** your back.

11. Hold one of your hands **beside** each ear.

12. Lay your hands **on top of** your shoulders.

13. Take your hands **off** your shoulders and put them **under** your chin.

14. Put one hand **below** the other hand.

15. Put your hands **next to** your sides.

16. Lift one foot and hold it **above** your other foot.

17. Then put it **on** the floor **beside** your other foot.

18. Sit **down** on your chair.

— Practical Application —

Distribute a box and an object to each student. Have them move their objects to positions in and around their boxes according to the following directions.

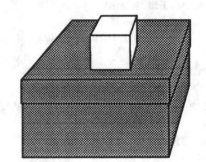

1. Put your (object) **on top of** the box.

2. Take it **off** the box and put it **beside** the box.

3. Hold your (object) **over** the box.

4. Hide your (object) **in back of** the box.

5. Move it to the **front of** the box.

6. Put it **inside** the box.

7. Take your (object) **out** of the box.

8. Put it **behind** the box.

9. Move your (object) to a place **above** the box.

10. Move it to a place **between** you and the box.

11. Then move your (object) to a place **next to** the box.

* * * * *

Play "Place in Space." Give a student a chalkboard eraser. Tell him/her to place it in any space around the room while the other students watch. Another student can get the eraser after describing its position (for example, under the table, on the floor between the door and the desk, behind the book, in the desk). This student then places the eraser in another space. Give as many students as possible a chance to place or describe the eraser's "place in space."

* * * * *

After students are comfortable with the Place in Space exercise, they can try a more complex version of the same activity. Have students, one at a time, visually select a classroom object, then describe its "place in space" without naming it. For example, "I spy an object that is on the wall next to the bulletin board, between the clock and the sink, and below the poster." The rest of the students then try to figure out which object occupies the space being described.

* * * * *

As an extending activity, have the class, as a group, write a language experience story using relative position words. Draw pictures to illustrate positions mentioned in the story. Have students make puppets of story characters, then use the puppets to act out the story. Also, call students' attention to illustrations in their reading and social studies books and have them describe the positions of characters and objects.

ACTIVITY 37: THE PATH OF LIGHT

Goal: To understand that light travels in straight paths

Skills: Observing, discussing, comparing, summarizing

Materials: Flashlight
For each student:
black rubber tubing, if possible (about 12" long, 1/2–1" in diameter)

Preparation: None

Preparation Time: None

Lesson Time: 20–25 minutes

— Procedure and Questioning Strategy —

Darken the classroom as much as possible so the beam from a flashlight can be seen more easily by the students. Hold the flashlight parallel to the floor and shine the light directly across the room at a blank space on a wall or on a chalkboard.

1. Where is the light source? Where is the light coming from?

   The light is coming out of the flashlight.

2. Look at the beam of light going across the room.

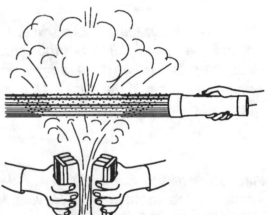

If the beam of light is not clear, have one of the students clap two chalkboard erasers in front of the flashlight so the chalk dust can help define the light beam.

3. Is the path of the light beam straight or bent?

   It's straight.

4. How do you know?

   We can see the light shining across the room on the wall (or chalkboard) and it goes straight across.

5. I am going to shine the light on different parts of the room. Observe the beam of light carefully as I move it around the room.

Shine the flashlight around the room without shining it on any of the students. Hold it at different levels and also hold it above your head and shine it straight down.

6. What do you observe about the beam of light? Is its path always straight or does it bend or curve?

   It's always straight.

Turn on the light in the classroom. Distribute a piece of black tubing to each student.

7. Choose a light source such as a window or the overhead lights on the ceiling. Look through one end of the tubing and point the other end toward your light source. Can you see the light source through the tubing? Try one eye at a time and try holding the tubing curved, bent, and straight.

Allow students enough time to experiment looking through their tubing at different light sources and with their tubing bent or curved in different directions.

8. When can you see the light?

   When I hold the tubing straight. When I point the tubing at the light source.

9. Is light coming from your eye?

   No.

10. Where is the light that you see in the tubing coming from?

    It's coming from the windows or the lights on the ceiling.

11. So the light you see in the tube is traveling from the light source through the tube to your eye.

    Yes.

12. How are you holding the tubing when you see the light?

    When I hold the tubing straight out.

13. Why can you only see the light traveling through the tubing when you hold the tubing out straight?

    Because light only travels in a straight path. It can't get through the tubing when the tubing is bent or curved.

— **Practical Application** —

1. Look around our classroom. Aside from the windows, where does the light come from?

   From the lights on the ceiling.

2. Why do we need lights in our classroom?

   So we can see to do our work.

3. When I held the flashlight above my head and pointed it down, in which direction did the light travel?

   It traveled down.

4. What part of the room did it light up?

   It lit up the floor.

5. Why did the light end up on the floor?

   Because it traveled in a straight path down from the flashlight.

6. So why do you think our classroom lights are on the ceiling and not on the floor?

   Because the light needs to shine straight down on the top of things.

ACTIVITY 38: BLOCKING THE PATH OF LIGHT

Goal: To understand that some materials block the path of light

Skills: Exploring, observing, describing, classifying, explaining

Materials: For each group of 4 students:
    flashlight
    cardboard tube from a paper towel roll—or any tube that will work with
      the flashlights used in this activity (Save these tubes for the next lesson.)
    sheet of plain white paper
   Materials to test to see if they will block light:
    pieces of waxed paper, tissue paper, aluminum foil, dark construction paper,
      clear plastic wrap, white paper, facial tissue, colored cellophane
   Rubber band

Preparation: Assemble the materials for each group.

Preparation Time: 5 minutes

Lesson Time: 20–25 minutes

— Procedure and Questioning Strategy —

Tape the sheet of white paper to a wall or the chalkboard where all students can see it. Turn on a flashlight and demonstrate how to hold a cardboard tube at the end of a flashlight so the light shines through the tube onto a sheet of white paper.

Darken the room so the light from the flashlight can be seen.

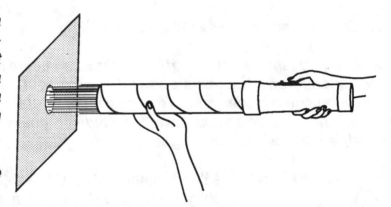

1. Look at the light shining on the paper. Where is the light coming from?
   The flashlight.

2. What shape is the light on the paper?
   It's a circle.

3. Look at the end of the tube. What shape is it?
   A circle.

Take the tube away from the flashlight and point the end of the tube toward the students.

4. Why do you think the light shining on the paper is in the shape of a circle?

> The tube has a circle shape.

Shine the flashlight through the tube at the sheet of paper again. Have a student hold a hand over the end of the tube.

5. The flashlight is still turned on. What do you notice about the sheet of paper?

   > The light from the flashlight doesn't show on the paper anymore.

6. Why do you think it's not there?

   > (Student's name) hand is in the way.

Have the student move his/her hand away from the tube so the other students can see the light on his/her hand.

7. What do you see now?

   > The circle of light is on (student's name) hand.

8. So we can say that the light was stopped, or blocked, by the hand. It couldn't go through the hand to the paper.

— Practical Application —

1. Let's find some other things that block light.

Divide the students into groups of four. Take a cardboard tube and a piece of waxed paper and demonstrate how to hold a piece of paper over the end of the tube with a rubber band. Then hold up each of the materials to be tested, one at a time, and have the students guess whether these materials will or will not block light.

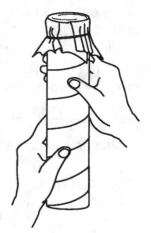

2. Each group will have these materials to test to see if they block the light from the flashlight. Try each kind of material. Hold it on the end of the tube with a rubber band. Then turn on your flashlight and see if you can see the light shining through the tube onto your sheet of white paper.

Distribute materials to each group of students. Allow enough time for students to explore with the materials and test their guesses.

3. Put the materials that block the light in one pile and those that do not in another pile.

Discuss which materials blocked light, which did not, and which materials did not fit into either group. Have the students explain why. Encourage them to look for similarities among materials in each pile and to relate their findings with whether or not the materials blocked light. Have them suggest other materials at school or at home that they think would fall into these categories.

ACTIVITY 39: INDOOR SHADOWS

Goal: To understand that shadows are made when an object blocks the light

Skills: Observing, describing, comparing, discussing, explaining, summarizing

Materials: For each group of 4 students:
  flashlight
  cardboard tube from a paper towel roll
  sheet of plain white paper
  4 index cards
  transparent tape
  scissors
  Filmstrip projector
  Large white bedsheet
  Classroom objects
  Stuffed animals brought in by the students (optional)

Preparation: 1. Assemble the materials for each group.
  2. Set up the filmstrip projector and hang the sheet on the wall.

Preparation Time: 10 minutes

Lesson Time: 30–35 minutes

— Procedure and Questioning Strategy —

1. Let's make some shapes to put on the ends of our cardboard tubes.

Demonstrate what to do as you give directions:

2. – Look at the size of the hole at the end of your paper towel tube. The shape you draw will need to be small enough to fit inside that circle.

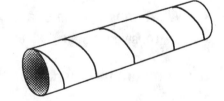

– Put the end of your paper towel tube down on an index card. Draw a circle around the end of the tube like this.

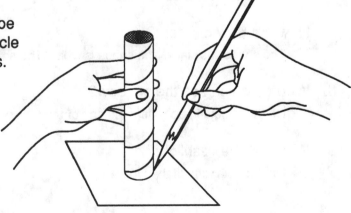

– Draw a shape that touches the top and bottom of the circle on your card, but not any of the other parts of the circle.

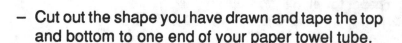

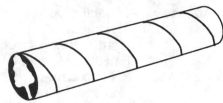

– Cut out the shape you have drawn and tape the top and bottom to one end of your paper towel tube.

3. After you have made a shape and attached it to your tube, hold your flashlight on the other end of the tube so it shines through the tube onto your sheet of white paper like this.

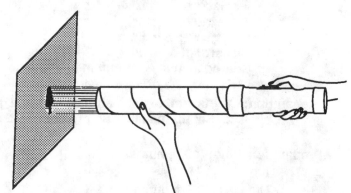

Divide the students into groups of four and distribute materials to each group. Tell the students that they will be sharing the flashlight. While the students are testing their shapes on the tubes, walk around among the groups. Listen to the students' comments and guide them to think further about their observations. Some guiding questions might be:

4. What shape is the shadow on your paper?

   (Answers vary.)

5. Compare the shadow on the paper to the shape at the end of the tube.

6. How are the shapes similar?

   They are both the same shape.

7. How are they different?

   The shadow is black. The shape is white. The shadow and the shape are different sizes.

8. Where do you see the shadow?

   On the sheet of paper near the end of the tube.

9. Where is the flashlight?

   On the other end of the tube.

10. Where is the shape that is making the shadow?

    (Guide the students to observe that the shape is between the light from the flashlight and the shadow on the paper.)

11. How did the shadow get on the paper?

    (Answers vary.)

12. The shape at the end of the tube actually stops or blocks the light so the shadow of the shape is black. The light that passes by the cut-out shape at the end of the tube outlines the shape.

When the students finish testing their shapes, discuss the results of their experiences. Ask the same kinds of questions that you asked the small groups of students so all students can share their findings. Guide the students to understand that in order to make a shadow, the object making the shadow has to be **between** *the shadow and the light source. Shapes that make shadows are made of materials that light cannot go through. Shadows are made by objects that block the light.*

— Practical Application —

Set up the filmstrip projector so it lights up as much of the bedsheet hanging on the wall as possible. Have different students hold up classroom objects between the light and the bedsheet and experiment with the position of the object in relation to the sheet.

1. Where is the shadow?

    On the sheet.

2. What shape is the shadow?

    (The same shape as the object being held up.)

3. How can you make the shape larger? Try it.

4. How did you do it?

    By moving the object farther from the sheet.

5. How can you make it smaller? Try it again.

6. How did you do it?

    By moving the object closer to the sheet.

7. When is the shadow clearer?

    When the shadow is smaller.

8. When is the shadow fuzzier?

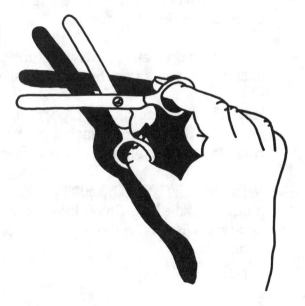

When it's bigger.

Have a student hold an object behind the light from the projector.

9. Can you see a shadow?

   No.

10. Where is the object?

    Behind the light.

11. Where does the object need to be to make a shadow?

    Between the light and the sheet.

12. What makes the shadow?

    The object blocks (or stops) the light.

Have as many students as time permits make shadows with classroom objects or stuffed animals.

13. Can you make the shadow dance?

14. Can you make the shadow a different shape?

    (Have them hold their objects sideways, turn them around, and turn them upside down.)

* * * * *

— Guide the students to create a shadow puppet show. Puppets can be cut out of paper and glued on tongue depressors.

— Even more dramatic . . . . Have students put on a shadow play so their own shadows are cast on the sheet from behind it. Make up the play together.

— Have the students cut out larger shapes and put them on the overhead projector so they project on the screen. What happens when the shape is turned over — or turned around? How does that change the shadow on the screen?

— Make shadow boxes by cutting shapes from paper and gluing them so they stand up in on open shoe box. Shine the flashlight into the shadow box. Where do the shadows appear? What if the flashlight position is changed? Do the shadows appear longer or shorter?

ACTIVITY 40: OUTDOOR SHADOWS

Goal: To apply the understanding that shadows are made by blocking light

Skills: Observing, describing, comparing, discussing, explaining

Materials: Two very large pieces of paper for tracing students' shadows
Crayon or marker (dark color)

Preparation: None

Preparation Time: None

Lesson Time: 25–30 minutes

— Procedure and Questioning Strategy —

Start this activity early in the morning on a sunny day when shadows are the longest.

1. We have made shadows indoors by using a light, an object, and something to show the shadow on, like a piece of paper and a sheet. Where did an object have to be to make a shadow?

    Between the light and the paper or sheet.

If the students do not state this position of the object, model the position with a flashlight, an object, and a sheet of paper.

2. Have you ever seen shadows outdoors?

    Yes.

3. What kinds of things make shadows outdoors?

    (Possible answers: trees, bushes, fences, people, buildings)

4. We are going to go outdoors to find shadows. Then we're going to make some shadows of our own. **Do not look directly at the sun because it will hurt your eyes.**

Take the students on a walk outside to observe different shadows. Take the paper and the crayon or marker. While the students are observing shadows, ask them guiding questions.

5. What shadows do you see?

    (Answers vary.)

6. What made the shadows?

    (Answers vary.)

7. In which direction do the shadows go?

   (Students should notice that all the shadows go in the same direction.)

8. Stand in the shadow of a tree (or a building). We usually call this kind of shadow *shade*. Where is your shadow?

   I can't see my shadow.

9. Go back into the sun. Where is your shadow?

   Next to me.

10. Where is the light coming from to make your shadow?

    The sun.

11. Where do you see your shadow?

    On the ground. On the building.

12. How is your shadow made?

    By standing between the sun and the ground. By blocking the light with my body.

13. Stand with your back toward the sun. Where is your shadow?

    In front of me.

14. Where is the light source?

    Behind me.

15. See if you can change the size and shape of your shadow. Can you make it fatter? Can you make it skinnier?

16. Turn around. Where is your shadow?

    Behind me.

17. Point to your shadow and to the sun. Where are you in relation to these two things?

    Between them.

18. So why isn't your shadow still in front of you?

    Because even though I turned around, my body is still between the sun and my shadow.

Guide students to understand that a shadow will always be on the side of them that is opposite the sun or light source.

— Practical Application —

Pick out a shadow from a particular tree (or building). Starting at the trunk of the tree (or the wall of the building), pace off the distance from there, the bottom of the shadow, to the top of the shadow. Record the length of the shadow. Have students discuss the direction the shadow is going and find some object the shadow is pointed toward as a reference. Ask the students whether the shadow is as tall, taller, or shorter than the object?

Now put a large piece of paper on the ground away from other shadows so the full shadow of one student is projected onto the paper. Outline the shadow of the student on the paper. Discuss with the students the parts of the body that relate with the parts of the shadow.

* * * * *

Take the students back outdoors around noontime. Bring another large piece of paper and a crayon or marker. Again pace off the distance from the bottom of the shadow to the top of the shadow. Record the length of the shadow cast by the same tree (or building) measured that morning. Compare the lengths of the morning and noon shadows.

1. Where was the sun in the morning?

2. Where is the sun now?

Remind students not to look directly at the sun.

3. In which direction is the shadow going? Is it in the same direction as this morning?

4. When does the shadow move to the other side?

Outline the shadow of the same student on the piece of paper. When back in the classroom, have the students compare the drawings of the shadows made in the morning and at noon.

5. How is the drawing of the shadow we made in the morning different from the one we made at noon?

   The shadow we made at noon is smaller or shorter.

6. Why do you think the sizes are different? Was the sun in the same place at noon as it was in the morning?

   At noon the sun is overhead more than it is in the morning.

* * * * *

Read Robert Lewis Stevenson's "I Have A Little Shadow."

ACTIVITY 41: MIRROR IMAGES

Goal: To understand that images can be reflected from shiny surfaces

Skills: Exploring, observing, describing, comparing, discussing, explaining

Materials: For each student:
      plastic mirror (if unavailable locally, these can be ordered from a
         school science supply catalog)
      penny
      masking tape
      small colorful objects like shells, real and straw flowers, crayons, small
         pieces of print fabric, and small colored pictures from magazines

Preparation: Gather the materials and place them on a table where they can be
      easily reached.

Preparation Time: 5 minutes

Lesson Time: 25–30 minutes

— Procedure and Questioning Strategy —

Distribute mirrors to the students and give them time to explore with them. Walk around among the students and ask guiding questions while they are exploring. For example:
– Can you see your face in the mirror?
– Where are you holding the mirror to see it?
– Can you turn the mirror so you can see your knees? legs? feet?
– How are you holding the mirror?
– Can you turn the mirror so you can see someone else in the classroom?
– How are you holding the mirror this time?

Take some time for students to share what they have discovered while using the mirrors. Then divide the students into pairs.

1. Where do you need to hold your mirror so you can see your face?

    In front of my face.

2. There are two people in each of your pairs. You need just one mirror to try this, so put down one of your mirrors.

3. Stand about a foot apart facing forward. One person in each pair hold the mirror up to about face level. Move the mirror around until you can see each other's face in the mirror.

As the students are trying this, walk among the students and guide them to try different mirror positions. When they hold the mirror straight in front of them, between both of them, the students can see each other's image.

4. Where did you hold your mirror when you saw your partner's face in the mirror?

    In front of us in the middle. Between us.

5. When you could see your partner's face in the mirror, what could your partner see?

    My face.

6. So you both could see each other in the mirror at the same time?

    Yes.

Distribute a penny and a piece of masking tape to each pair of students. Demonstrate how to connect two mirrors by taping them together on the back with masking tape. Show how to hold two mirrors on the table, at about a 90° (⌐) angle, mirror side toward you.

7. Hold your mirrors together, on the table, like this. Put a penny on the table where the mirrors meet in the center. What do you see?

    Four pennies.

8. How many pennies did you put on the table?

    One.

9. How many pennies do you see reflected in each mirror?

    One, and half (or part) of another.

10. Slowly move the mirrors closer together. How does it change the number of reflections you see of the pennies in the mirrors?

    (Answers vary.)

11. Slowly move the mirrors farther apart. How many reflections do you see?

    (Answers vary.)

12. Hold your mirrors in front of you so they open and close up and down like a mouth. With your eyes close to the open part of the "mouth", look into the two mirrors. What do you see?

    Lots of reflections of my eyes.

13. Do all of your pairs of eyes look the same in the mirrors?

    No. Some pairs are right side up and some are upside down.

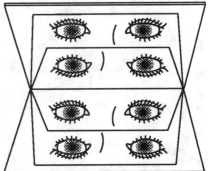

14. Which are upside down?

> Every other pair of eyes. The bottom pair in the top mirror and the top pair in the bottom mirror.

15. Close the mirror up a little. What happens to the reflections?

> (Answers vary.)

16. Slowly open the mirror wider and wider. What happens to the reflections?

> (Answers vary.)

— Practical Application —

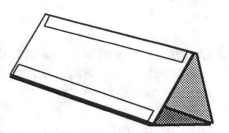

Collect the mirrors from the students and make several taleidoscopes (a kind of kaleidoscope) with mirrors and masking tape. Hold three mirrors together with the longer sides of the mirrors touching. The three mirrors will form a triangle. Have a student tape the mirrors in this position while you hold the three mirrors together. Put more tape around the taleidoscope to make it more durable.

Demonstrate how to hold the taleidoscope over something small like a shell, flower, or other small object. (You can see many reflections of the object in the mirrors.) Put the taleidoscopes, along with a collection of small objects, on a table so students can get a chance to look at them through the taleidoscopes. Suggest that they try looking at the fabric and print of their clothes — and other objects in the classroom — through the taleidoscope. Which were the most interesting to look at? How many reflections of an object were seen in the taleidoscope?

ACTIVITIES 36–41: CONNECTIONS

Goal: To find relationships among the activities for space, shadows, and light

Skills: Observing, comparing, classifying, discussing

Materials: Piece of white or light colored poster board
Dowel rod about 2 feet long (available at most hardware or building supply stores)

— Procedure and Questioning Strategy —

This activity needs to be started as early in the school day as possible.

1. **What have we found out about light?**

   It travels in straight paths (lines). It goes through some materials and not others.

2. **How did we make shadows?**

   By holding something in front of a light. By standing in the sun.

3. **How did doing that make a shadow?**

   It blocked the light.

4. **We found that the shadows we saw outdoors changed. They were shorter at noon than in the morning. Why did that happen?**

   The sun moved.

5. **Let's try an experiment to see how outdoor shadows change at different times during the day.**

Show students the dowel rod.

6. We'll put a piece of poster board outdoors on the ground. Then we'll stick this dowel rod through the middle of the poster board and into the ground. We'll go outdoors at different times during the day to outline the shadow of the rod on the poster board. Each time we do this we'll also write down the time we outlined the shadow.

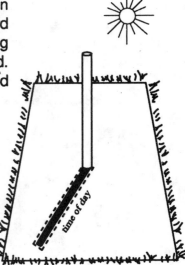

Have the students go outside with you when you set up the experiment. Have a student outline the shadow of the dowel rod on the poster board and write the time of day next to the shadow. Have the students glance very quickly and carefully at the sun to see its position.

Continue outlining the changing shadow, recording the time, and observing the sun's position at intervals during the day (each hour, if possible). Discuss the direction in which the sun is moving and the direction of the dowel's shadow. After the last observation, bring the poster board and dowel rod back into the classroom.

7. What made the shadows?

   The dowel rod.

8. How did it make the shadow?

   It blocked the light of the sun. It was between the sun and the poster board.

9. What happened to the lengths of the shadows?

   They changed during the day.

10. When were they the longest?

    The first and last times we outlined them.

11. When was the shadow the shortest?

    (It should be around noontime.)

12. What else do you notice about the shadows?

    They are in different places around the dowel rod.

13. Why do you think the shadows are different lengths and are in different places?

    The sun was in different places. The sun moved across the sky.

If students do not relate the position of the sun with the changes in the shadows, demonstrate this with a flashlight, the dowel rod, and poster board.